FLY FISHING

By Fennel Hudson:

A MEANINGFUL LIFE
A WATERSIDE YEAR
A WRITER'S YEAR
WILD CARP
FLY FISHING
TRADITIONAL ANGLING
THE QUIET FIELDS
FINE THINGS
A GARDENER'S YEAR
THE LIGHTER SIDE
FRIENDSHIP
NATURE ESCAPE
BOOK OF SECRETS
THE PURSUIT OF LIFE

Fennel's Journal

No. 5

FLY FISHING

By

Fennel Hudson

2019
FENNEL'S PRIORY LIMITED

Published by Fennel's Priory Limited

www.fennelspriory.com

First shared as handwritten letters in 2009
Limited edition magazine published in 2012
eBook published in 2013
This extended edition published in 2019

Copyright © Fennel Hudson 2009, 2019

Fennel Hudson has asserted his right under the
Copyright, Designs and Patents Act 1988
to be identified as the author of this work.

All rights reserved. No part of this publication may be
reproduced, stored in a retrieval system or transmitted,
in any form or by any means, electronic, mechanical,
photocopying, recording or otherwise, without
the prior permission of Fennel's Priory Limited.

"Stop – Unplug – Escape – Enjoy"
and The Priory Flower logo are registered trademarks.

A CIP catalogue record for this book
is available from the British Library.

Hardback ISBN 978-1-909947-18-4
Paperback ISBN 978-1-909947-19-1
Kindle ISBN 978-1-909947-16-0
Audiobook ISBN 978-1-909947-90-0

Designed and typeset in 12pt Adobe Garamond Premier Pro.
Produced in England by Fennel's Priory Limited.

CONTENTS

Introduction. 1
A New Beginning. 7
Thinking Like a Fly-fisher 11
Living the Dream. 17
All Laid Out. 23
A New Fly Rod 31
A Very English Fly-fisher 39
If Only Every Day Could Be Like This. 45
The Amwell Magna Fishery 49
Local Fishing 57
Fishing, Far Away. 63
A Duffer in May. 69
A Trout, from the Kidderminster? 81
Trout from the Hills. .91
Rewilding Fly-fishing. 97
Trout at Two Thousand Feet. 107
Alone but not Lonely. 117
The Net on our Backs. 125
A Grilling for a Trout. 133
Catch Returns 139
Looking Backwards, Going Forward 147
Grayling to Remember. 155
Fingers and Fluff 161
The Four Rules of Angling 169
About the Author. .171
The Fennel's Journal Series 175

STOP – UNPLUG – ESCAPE – ENJOY

This book, and the series to which it belongs, is about freedom. It's also about the adventures to be had when pursuing one's dreams, developing and communicating one's self, and striving for a slow-paced rural life. It's your opportunity to take time out from the stresses of modern living, to stop the wheels for a while, unplug from the daily grind, escape to a quiet and peaceful place, and enjoy the simple life. Because of this, I'd like you to read it in a distraction-free and relaxing environment: your 'safe place' where you can savour quality time and, if possible, delight in the beauty of the countryside.

That's why the book is pocket-sized, has a waxy cover, and is printed using a special waterproof ink. It's designed to be taken with you on your travels. Don't store it in pristine condition upon a bookshelf; allow it to reflect the adventures you've had. Use a leaf as a bookmark and annotate the pages in the spaces provided with ideas of how you will honour your right to 'never do anything that offends your soul'.

The more mud-splattered, grass-stained, and pencil-scribbled this book becomes, the more you've demonstrated your ability to pursue a contented country life. So go on: live your life, be authentic, and always remember to 'Stop – Unplug – Escape – Enjoy'.

fennel

*"I wanted to fish and fish in peace.
The dry fly I was using was a tenet of my faith."*

Ian Niall

INTRODUCTION

Fly-fishing: it's the ultimate form of angling. Each cast paints a line in the sky which we follow to trace the thread of our existence. From Primitive Man's attempts to feed his family, to the modern-day recreation that soothes our soul, fly-fishing is an act of hope that leads to a net full of fish and a head full of dreams. It's poetry in motion, albeit with a strict "Eleven o'clock to one o'clock" movement.

Fly-fishing, in this book at least, aspires to be an art form. Like many works of art, it is personal, subjective, emotive and heartfelt. It's designed to be studied and appreciated, but mostly to be enjoyed and loved. It leads to a connection with a fish that is often closer than in other forms of angling; it encourages a holistic appreciation of the watery environment and a courtship with the natural world. All the flora and fauna, the seasonal changes reflected by the water, the bedrock and land formations that influence the habit and health of rivers and lakes, and even the cycle and quality of water from raindrop to ocean and back again, they all influence the fish and, directly, the angler's sport.

Like all freshwater anglers, fly-fishers are custodians of water quality, passionately protecting the health of our rivers, streams, lakes, reservoirs, lochs and llyns. And as their casts aim seaward, so too are they seeking to protect the quality of our marine environments.

Angler-environmentalist? Fly-fishers are sometimes referred to as 'angler-naturalists' because their desire to understand the fish in its natural habitat, to know its diet at any time of day or year and deceive it with an artificial representation of its natural food, takes them into the fishes' world to connect with it on its terms. It is the opposite of modern-day bait fishing where anglers educate the fish to eat an unnatural bait. Both, though, are acts of 'angler versus fish'; sometimes to catch food for the table, sometimes for the pleasure of hooking, playing and returning a fish, and sometimes simply to prove that Man – as apex predator – is smarter and wilier than a comparatively small-brained aquatic creature. Alas, the fish often wins the day and the angler has to go home to nurse a broken ego.

Ego-challenge or not, fly-fishing is the simplest, purest, most skilful and pleasurable way to catch a fish. It's basic fishing, almost primitive in its simplicity. Herein is its beauty. The angler, unencumbered with superfluous tackle and tactics that burden his or her bait fishing brethren, carries just one rod, a line, reel, net, and collection of artificial flies. This 'lightness of intent' enables him or her to travel lightly and explore deeper

INTRODUCTION

into wild places. Leaving all the clutter behind, to savour the freedom of isolation and the companionship of the wild, replenishes the angler's soul. With a fly rod, anglers are not casting to a fish; rather to a circle of dreams: ripples that spread into every aspect of their lives. It is angling distilled into the pinnacle of sporting style and grace.

Let's think about that for a moment: *style and grace*. An outsider to the sport might find this description unusual for an activity where someone ties feathers, fur and tinsel to a hook and then flings it into the water in the hope that a fish will be fooled into to taking it into its mouth. Stylish? Yes, when the casts are smooth and the line flies true and straight. But graceful? It *is* a grace-full act, elegant at times and often prayerful, and a very refined form of fishing it is, too.

A fisherman catches, or seeks to catch, fish; often with a net. An angler is a subset of this where the use of a hook, otherwise known as an angle, is used to catch one's quarry. This is what separates fishing from angling. Angling, thus, is the art form; it's about deceiving a fish by subtle means. Fishing, on the other hand, is the more 'industrial' harvesting of fish, often trawling and scooping the fish from the water. Yet we don't say "fly-angling"; we say "fly-fishing". The hook, therefore, becomes a curve of steel that bends the description of fishing in favour of alliteration. *"Fly fishing"* sounds better. Simple as that. Indeed, fly fishing is founded on simplicity.

Not all of angling is simple. In fact, history is full of the debates that have raged between different doctrines. Anglers, it seems, are quick to distinguish and categorise types of angling, angler, and fish – grading them in terms of style, grace, sporting appeal and social class. They will say that those who fly-fish – typically for trout, sea trout and salmon – are 'game' anglers (these fish being edible and therefore 'fair game') and those who pursue these fish are in some way superior to those who do not. It's angling elitism. Aspirational, yes, but rather antagonistic and snobbish. Those who angle for other species of freshwater fish are deemed to be 'coarse' anglers, because the fish they pursue are reputed to be less edible, being coarse and vulgar to the taste. (This may be true for some bottom-feeding fish, that suck their food from the mud at the base of lakes and slow-moving rivers; but predators such as perch and zander make fine eating.) The reality, of course, is that such divides are nonsense. All fish are born equal. It's we who encourage the prejudice – both in terms of their categorisation and the tactics employed to catch them. If an angler chooses to fish for carp using a fly, or salmon using a worm, then they are simply displaying personal preference. It's art, remember; a subjectively beautiful and deeply personal thing. There is, however, such thing as a coarse angler. Some anglers, be they bait *or* fly fishers, lack social and sporting grace. Often, they are so

INTRODUCTION

focused on the act of catching fish that they are unable to fully appreciate the joys of angling or be sensitive to those around them. They're obsessed with the end result rather than the leisurely process that occasionally brings a fish to one's hand. They can be efficient harvesters of fish, for sure, but as they curse and rant, they're forgetting that angling is merely a recreation. It's meant to be leisurely, peaceful and relaxing. Those who seek to empty a water of its fish might as well be using dynamite to blast their 'quarry' into submission. Their seriousness, competitiveness, urgency, self-imposed pressure to perform, and apparently insatiable hunger (just how many fish does someone need to catch, kill or eat?) tell us that they've missed the point of angling. This book is not for them. Instead, it's for those who seek quiet times by the water and know that angling is a gentle, romantic, pleasurable and quirky sport practised in harmony with nature. It's the *experience* that counts.

Fly-fishing is a timeless act, where one can experience the same drama and emotions as one's ancestors. To become an angler, therefore, is to continue a tradition that's fed our bodies and souls since Man learned to hunt. Yet many anglers do not know where their calling (or hunter's instinct) will lead them. They are content to drift with the current, so to speak, along a watery path that will evolve and grow much as a river widens with age. They will have fleeting fancies, dalliances with different species and waters. But for some, angling enables a quest into the unknown, to discover wild secret places. For these fortunate people, fly-fishing is escapism. Their hours by water serve as contemplation that enriches their souls, directing their quest inwards, to achieve a longed-for state of completeness. It's where they can slow their casts to mirror the relaxed rhythm of life. Maybe, if they're lucky, they'll realise that they're connecting with what's important. This is what this book seeks to achieve: a slowing down of the otherwise frantic pace of life, to marvel at the wonders of water and the fish it contains. Its chapters are written monthly, to capture the changing seasons and perspectives through the year. By the time you get to May, I'm hoping that you'll have allowed me to reel you in to my way of thinking. There, in playful mood, we will celebrate the riches, joys, idiosyncrasies, spectacle and humour of a fly-fishing life before we journey to deeper waters and the truths they contain.

I hope you enjoy the adventure.

JANUARY

I

A NEW BEGINNING

I am an angler. A fly-fisher. I cast in hope, attempting to capture my dreams. Doing so defines me, much as being a writer shapes my identity and purpose. Perhaps you are a fly-fisher, too, and can relate to what I'm saying? Or maybe you're curious to know more about this finessed form of angling, where a fly rod is compared to a wand and a piece of fluff tied to a hook can be revered as a work of art? To know why something is so important, it pays to go back to the beginning…

Have you ever pondered upon the things that most influenced your life? Perhaps it was an event, or a person, or a crossroads that encouraged you to try something new? Maybe it was an opportunity to reflect and redirect your thinking or actions, when your learning kicked in and you appreciated the value of memory and dreams? I'm having one such moment right now. There's nothing especially urgent for me to do, so I have the opportunity to browse through my bookshelves and read a few chapters from my favourite angling books. I've read a few pages of John Gierach's *Even Brook Trout Get the Blues* and Harry Plunket-Greene's *Where the Bright*

Waters Meet, but have since found myself pleasantly distracted by some photo albums that also live on the bookshelves. The first album I viewed contains photos from my college years. There am I in flat cap, tweed shirt, corduroy trousers, braces, fingerless mittens, and hair that looks worryingly like a Dolly Parton wig. A strange look, for sure, but I was an art student expressing myself in a creative way. Really, though, I was searching for a way to find myself – even though I was already found. Caring not for usual teenage pleasures, I was trying to work out how to fit in to a world that seemed to involve spending too much time with too many people – and not enough time fishing. This presented many challenges. A flick through earlier photos revealed why I struggled with crowded places and the bustle and noise of urban environments.

Early photos of me document a childhood spent beside water. There are plenty of pictures of me holding a fish, or fishing, or searching for somewhere to fish. Interestingly, the fish in these photos morph over time at the same speed as my hair grew long and my clothes became shabby. They begin as wild brown trout from mountain streams, and then become fatter and smaller-finned stocked varieties during my teenage years. All of the fishing was fun, but the fishing at the beginning was the most authentic.

The beginning. That's what's it's about. The source of one's identity and passions. There, at the front of my first photo album, is a picture that documents the origin of my love of angling: it shows me cradling my first fish – a little brown trout that was caught from a mountain lake in Wales. The year was 1980 and I was just a few days short of my sixth birthday. I remember the exhilaration as the fish took hold of my Devon Minnow spinner, the panic I felt as it fought for freedom, and my overwhelming sense of elation when the fish was safely in the net. This moment in my life, which lasted no more than two minutes, was to define the years that followed. I had proudly become an angler.

At the age of eight, I picked up a fly rod and attempted to cast. It was such a difficult task. The rod and line felt impossibly heavy, the timing and energy of the forward and backward casts – which needed to be slower and more powerful as the line extended – was almost impossible to grasp, resulting in heaps of line at my feet or whip-like cracking in the air. I wondered why I should bother with something so difficult compared to casting a spinner with a short rod and fixed spool reel. But I'd seen my father and uncle fly fishing with apparent ease – and catch more fish than everyone else – so fly fishing had to be the secret to success. So I stuck with it. Eventually, following much practising on the garden lawn, I began to 'feel' the loading of the rod and build the muscle memory for the correct timing. Casting became easier and more enjoyable. Then, for my 13th birthday, I was presented with my very own fly rod.

Dad had acknowledged my casting abilities and was welcoming me to the fold. I was now a bona fide fly-fisher. And I was exclusively so. There would be no more hurling of spinners. I had my new means, and terms, of fishing. Indeed, every time I picked up that fly rod, I knew that I was part of a global brethren of fly-fishers. It felt good then, it feels great now. I am first and foremost a fly-fisher. But I'm not an angling snob. I know that all fish are equal and it's the richness, diversity and seasonality of angling that makes it eternally appealing. Hence why, from my late teens onwards, I have fished for all freshwater species and use bait and fly techniques. But I was always *fundamentally* a fly-fisher.

For the duration of this book, fishing shall only involve hand-tied flies. And a fish, for the purpose of our vision, shall principally be a trout; though grayling will also be pursued and salmon and sea-trout will be mentioned. At the 'grand young' age of 35, I'm rebooting my angling, returning to my roots, to capture the same image as that photo of me with my first fish.

This is it: my – indeed our – new beginning. One where trout have risen to the surface, flies are being gracefully cast through the air, and salmon are battling the torrents of spate-filled rivers. And so, as I stare at the photo of me with my first fish, and think about past, present and future, I raise a toast to "New Beginnings."

JANUARY

II

THINKING LIKE A FLY-FISHER

"Pay attention! It's nearly time! I should have known you'd be the one causing me issues. Now remember: focus on the part. Put yourself inside the mind of the character. See and feel as he does. Know that this is the greatest moment in his life and that he is about to become part of history. Never before has his kind been elevated to such greatness or been revered by so many. Lift your chin, straighten your back, feel the words forming in your mouth. They are rounded and loud. Rounded! And loud! Go my child! Take to the stage! Play the part you were born to perform! Cue…Fennel!"

"Baaa. Baaaaaa!!"

"Splendid! You are the best nativity sheep we've ever had…"

Being four years old and thrust onto a stage wearing nothing but black leggings and a roll of cotton wool was, shall we say, 'character building'. My schoolmates had landed trendier parts, such as Wise Men or Reebok-wearing donkeys. But not me. I was the one prancing around the stage like an oversized cotton-bud queuing for the toilet. I felt about as comfortable as a budgie on stilts and just as likely to topple to the ground for fear of being seen.

'Baaaa' indeed. I've never recovered from the experience, but what it taught me was that if I'm going to play a part, or identify myself with something, then I need to get inside the head of the character and understand why he or she behaves as they do.

Given that my thoughts at the moment are of fishing with a fly, I thought I'd repeat the experience and get inside the head of he who would fish in this way. Let's consider what it means to be a fly-fisher.

When you think of fly-fishing, do you picture the rhythmic art of casting a fly line? Does that line land in a perfectly straight line, possibly with a little 'jud' on the reel as the line reaches its full extent? Does the fly land gently on the water and then, in a matter of seconds, the water around it swirl and the line tug violently? Did you strike and feel the jagging sensation of a hooked fish? Look up. Where are you fishing?

Fly-fishing, in Britain at least, conjures up a variety of scenes, from glass-like southern chalk streams fished by Halford and his ilk, to rock-strewn tumbling rivers beloved by rugged countrymen who don't mind if their fly remains permanently submerged. A range of streams, stillwaters, anglers, flies, and doctrines exist in between. Such is the breadth of fly-fishing and the almost religious beliefs it encourages. But let's not go too deep. Not just yet.

The greatest study into the psyche of angling and anglers is not, as you might expect, Izaak Walton's *Compleat Angler*. It is a similarly titled and far more poignant book entitled *Compleat Tangler* by cartoonist Norman Thelwell. Famed for his drawings of little dumpy children riding even dumpier horses, Thelwell was a master of capturing the character of a sport. His study of angling is a masterpiece. Why? He got it right. That's why. It's a tongue-in-cheek but respectful view of a sport he loved. And when he came to describing fly fishers, he rightly identified the distinctions within the ranks. There were dry fly purists, wet fly anglers, and salmon fishers, each drawn as perfect caricatures. He also identified the vast and more obvious differences between coarse, game and sea anglers.

Differences in angling? Hmm. Is it a case of 'never the twain shall meet?' or should the all-round angler simply choose a different hat when crossing the stile from one camp to another?

As I grow older, and this year especially (it is thirty years since I caught my first fish), I have greater affinity with the game angler. Perhaps I'm just getting longer in the cast and that fly-fishing suits my mellowing tastes? (It's true that I listen to more blues music than classic rock these days, and soon I'll be wearing oversized headphones and listening to *Oldgit FM*.) But with angling, I'm not just compelled to cast a fly, I'm also slightly repelled by modern bait fishing.

Bait fishing, or coarse fishing as it's called in the UK, relies more on entrapment than deception. 'Maggot drowning' as some call it, where the fish are educated to eat man-made meals such as bread, luncheon meat, and day-glow pink pineapple-and-cheese flavoured pastes. Such fish are caught when their greed or inquisitiveness is greatest. With fly-fishing, for wild trout at least, the art is to deceive the fish into taking a fly that it thinks is a natural insect (bonus points if you've tied the fly yourself). To do this effectively, the angler should study the natural diet of the fish at various times of the year and present the correct artificial fly to 'match the hatch'. It is angling based upon entomology and skill rather than bait recipes and science. Given this information, which title is more appealing: *The Fly Fisher Naturalist* or *Coarse Fish Baiter*? Do you want to get closer to nature when fishing, or change nature to suit the contents of your bait box?

Fly-fishing, it would seem, retains more of the basic art of angling than modern bait fishing. With fly-fishing, tackle and tactics remain simple. It is a hunter's craft, where the quarry is stalked. The bait fisher, on the other hand, will often encourage the fish to come to him. He will lay the fishes' dinner table and then settle down to await his guests. This less mobile approach encourages him to carry all manner of superfluous home comforts, such as an umbrella, deckchair, tent, radio, refrigerator, lounge slippers, portable Jacuzzi, and wide-screen television. You get the picture? The fly fisher, on the other hand, travels light and remains mobile. He or she goes to the fish in search of bites and adventures unknown to the static bait fisher.

Someone once told me (perhaps it was Thelwell, in his book, or my faith healer down the pub?) that to be a fly fisher, at least a dry fly man, one has to develop a resolutely stiff upper lip, have a bulging wallet and a desire to eat one's catch as soon as it is lifted from the water. This isn't entirely true. Most fly-fishers are everyday folk. But there is an air of sportsmanship to fly-fishing that elevates it above the bait dangling of the masses. I like this. It makes me feel special, as if I'm stepping into a private jet to drink champagne with the super-elite. (Where, once I'm savouring the dizzy heights of angling's master race, I shall look down upon my distant bait fishing relatives and frown, asking them the simple question: "Why?") Thankfully, I have a sense of humour and a tongue pressed *firmly* into my cheek. I am too cack-handed a fly fisher to aspire to such snobbery. Besides, a fish is a fish. The humble gudgeon has just as much right to be in the river as a salmon. How you fish for it is merely a matter of personal preference. (Mind you, I'm not likely to be hosting the '*Gudgeon On The Fly*' Anglers Group Social anytime soon. I don't imagine many anglers would sign up to GOTFAGS?) So I'll maintain my poetic and slightly idealistic perspective, while remaining firmly grounded. A good choice, given that my head is usually in the clouds…

February

III

LIVING THE DREAM

It is said that we anglers are dreamers, and that it's the dream that drives us to seek out the perfect angling experience. Living the dream, at least the dream of this angler, requires two things: first is a happy home life, second is the perfect place and way in which to fish.

The dream of a contented, balanced life sounds achievable when things are good: when angling is the icing on an already rich and fulfilling cake. But what about when life is imbalanced? When we need to rely more on our time by water to make us happy? This is what happened to me two years ago, when I camped by a lake and wasn't seen by 'normal folk' for nine months. I learned a lot, mostly about myself; realising that it's easy to lose sight of what's important, and that the role angling plays in one's life is to complement not define.

If you've read the book, or seen the film, called *A River Runs Through It* by Norman Maclean then you'll know what I'm talking about when I describe a life with family and fishing at its core. Consider such a life: it is a Sunday; you wake, kiss your wife and cuddle your children; you make and eat breakfast then walk as a family to church where you renew your faith; you return home to cook and eat a roast dinner. Afterwards you relax in your armchair and then, when your children are playing and your wife is happily reading, you decide that – as everything is perfect – you will go fishing. You collect your tackle and walk or cycle to the river. You fish, caring not whether you catch, because you are happy just to be there. And you're just as happy returning home as you were to leave. Your evening is then spent doing whatever the family chooses; and when you go to bed contented you think of Maclean's words that "life every now and then becomes literature...as if life had been made and not happened." We are, after all, authors of our life story.

After ten years of hope, hard work and determination, I am closer to achieving my dream of a contented rural life. I have said farewell to my native Worcestershire and moved to the Cotswolds, where the lovely River Windrush flows through meadows next to my home. The river is central to my new life. It provides my local fishing, favourite walks, and nature studies. It's where trout rise, deer graze and drink, water voles and kingfishers make their homes and, love them or hate them, otters feed each morning.

The Windrush is a friendly river. As H.R Jukes once said of his *Loved River*, "it is very little wider, and just as winding, just as flower-strewn and fragrant as a little country lane. And just as gossipy." On those occasions when I can stand beside its flow, when everything except the river is still and sunlight is yet to command the day, I can call to it without speaking and know that it is 'my' river. It's my dawn pleasure and afternoon adventure. All within a hundred yards of home.

My environment wasn't always like this.

I grew up in a village on the outskirts of the industrial 'Black Country' area of Birmingham. It was a quiet and out of the way sort of place where, up until the turn of the twentieth century, some of the residents still lived in caves. These 'rock houses' eventually became the playground for local children who, when they weren't pretending to be troglodytes, would sit by the local river and guess what colour it would turn next with the hourly influx of ore from the smelting factories or dye from the carpet factories upstream. The river was the Stour. Not the delightful, clear-watered Stour of Dorset, but the West Midlands Stour that flows through all manner of industrial grime. It heaved and frothed its way through brick and concrete until it met open countryside and was eventually washed clean by the River Severn at Stourport.

The West Midlands Stour ('Stewer', as we pronounced it) was not the sort of place to inspire a boy to become an angler. It held as many fish as a festering spittoon and was little more than an open drain for the soot-blackened factories upstream. Its waters would be coloured purple, red, orange, and yellow during the course of a day. So much aquatic life was lost to the multi-coloured poison. The potential nirvana of having a fishing river so close to home, therefore, was not to be. I mention the River Stour not as an example of river degradation but of the potential joy of having fishable water so close to home. (How nice the dream would be to gaze from one's window, to see trout rising lazily in the afternoon sun, and know that all we'd need to do would be to grab our favourite fly rod and walk to the end of the garden.)

When I was younger, and being influenced by the misfortunes of the Stour, I was convinced that rivers further afield must provide better quality fishing. I drove hundreds of miles to the rivers of Hampshire and Dorset, ignoring countless delightful streams, rivers and ponds along the way. Water, or so I thought, was bluer next to the horizon. *Such is the fool who attempts to cast furthest.*

It was only when I moved from the village to a town (in search of work) that I realised I had taken the Stour for granted. Instead of waking to the calls of moorhens and herons, I was greeted by the drone of traffic and a silence that fills meaningless conversation. I vowed that someday, when my fortunes had improved and I was in a better position to govern my destiny, I would move from the town and purchase a property next to a river. It would be a beautiful river, clear-watered and pulsing with life, where mayflies flutter in spring and where trout oblige the local angler. It would also be a special place, one that would inspire me to keep a fly rod by the kitchen door and an angling diary at my side.

Mrs H and I debated over where exactly we wanted to live. I wanted a quaint and historic cottage with no visible neighbours and she knew what we could afford. So we compromised, buying a new house built where the famous Witney blankets used to be made. It's a temporary abode for us, at least until we can save enough money to get a proper cottage, one worthy of the Priory name. But the river flows nearby and, when I see mayflies rise from the river, I'll know that I'm so very close to realising my dream.

Windrush: the river that 'meanders through the rushes'. I'm gazing at it now, from a bridge that passes over the millstream next to my house. I'm looking across the water meadow, where buttercups are flowering and swallows are swooping to catch their dinner. A line of willows marks the course of the river that tumbles over gravels and then flows quietly before roaring over a weir and into the millstream. Soon it will be time for me to rest this pen, grab a fly rod and then walk along the river to see if I can stalk a trout for tomorrow's breakfast. But before I do, I need to declare something: *I love fly-fishing*, especially when it's local. An hour before breakfast, or two after work, does not impact other commitments. It enables me to enjoy a balanced life, stay connected, and live the angler's dream.

Postscript:

The West Midlands Stour has been saved by The Environment Agency. They've stopped the polluters, cleaned up the river, and given hope to the anglers of the Stour valley. Insect life has returned and trout have recolonized the river. And, with great celebration, salmon were seen leaping the Stour's weirs this spring. Maybe I'll return to fish the Stour? Maybe I'll catch a trout? Maybe I'll cast a fly in the town? Or maybe I won't. For now, the river I can see between the willows is not the Stour; it is the Windrush. Beautiful and enchanting, always close, always welcoming, always there, for when I need it.

March

IV

ALL LAID OUT

Choosing to become a fly-fisher, or at least committing to being a fly-only angler, makes me feel like I'm washing away the grime from my fishing tackle to reveal the beauty of what lies beneath. Things look refreshingly bright when they've had a good clean, don't they? Reels shine, rods glisten and tweeds take on a contented but confused look, as if saying, "You're not going to make an honest jacket of me, are you?" My perspective of angling is similarly renewed. A walk along the river will see me studying the shallows rather than deep pools and my drive to work has become much riskier. (Whenever a bug hits the windscreen I end up examining it rather than washing it away with the wipers. "Ah, lacewing: nice. Mayfly: I ought to go fishing tonight. Cigarette butt: oh, blummin' litter bugs.")

Fly-fishing encourages us to dream – of rose-tinted sunsets and lazy spring days when swallows swoop and the hedgerows are blossomed in brilliant white. William Caine summed it up for me when he wrote: "My spirit winged its way through infinite leagues of space until I saw below me a pleasant valley where a clear stream meandered among the water meadows." This is where fly-fishing transports us – to a place of mind and to the clear waters where trout rise. So why, for the love of God and all that's beautiful in this world, am I sitting beside a water-filled hole with my hand over my eyes and my first clenched awkwardly around this pen?

Two weeks ago, I was sitting at work and dreaming of the start of the river trout season. (It begins hereabouts on the 1st of April.) A fly-fishing colleague came up to me and said, "Hey, Fennel, how about a day's stillwater trout fishing to get your arm in for the new season?"

The invitation sounded promising. "Okay," I replied, "where shall we go?"

"It's all arranged," he said, "I've booked us a corporate day at a specimen trout fishery. Just don't bring any of your usual Nancy-girl bamboo rods; the trout there are BIG so you'll need to tool up and get serious."

Tool up and get serious? It sounded like an invitation to the Male Strippers' Annual Chess Tournament. I didn't fancy the idea of wearing a posing pouch and standing beside a lake 'twiddling my pawn', so I decided that if I were to go, then I'd fish as I always do and hang the consequences. I agreed to the invitation and waited for the day to come.

I arrived at the fishery at nine o'clock. Wow, it was busy. The car park was already full of an impressive array of those huge and gleaming 'on road' 4x4 motorcars that never see mud but can, if their drivers are especially adventurous, mount curbs and park on gravel driveways. Next to the car park was a timber lodge complete with café, shop, toilets and even a purpose-built 'filleting room' (which, on closer inspection, housed a dustbin full of fish guts and carcasses of trout that looked like they'd been mauled by grizzly bears). If this wasn't alien enough to me, when I turned to look at the lake, I saw, coming towards me, an angler pushing a wheelbarrow. It contained two enormous trout. Their heads and tails were flopping over the sides of the barrow. They had the look of fish that had been promised a five-star break but had ended up in a youth hostel in Bognor.

"Finished already?" I asked the angler as he reached the car park.

"Yeah," he replied, "I had these on my first two casts; twenty-two and twenty-four pounds; a rainbow and a brown; I'm off back to the lodge to purchase another ticket."

Twenty-two and twenty-four? This was supposed to be trout fishing, not carp fishing. Forty-six pounds of trout caught in two casts? That's more weight of fish than most anglers catch in a year, yet he wants to catch more? *Ours is not to question why, but sit awhile then make a sigh.*

After introducing myself to the fishery manager and collecting my permit, I gathered my tackle from the car and walked down to the lake. There were thirty swims from which to fish and twenty-eight were already taken. My colleague opted to fish a vacant swim near to the lodge (he's never far from a bacon butty and a freshly-made mug of tea). I headed to the far end of the lake where, it seemed, the proximity of trees made casting difficult. Most of the anglers I passed on the way were casting long lines with unflinching accuracy. I say most, because one angler, bless him, was obviously new to fly-fishing and, with each challenging cast, grimaced like a man chewing a pineapple-flavoured bumble bee. The most expert-looking angler was wearing metallic pink sunglasses and a camouflaged T-Shirt. (A fashion *faux pas*, for sure, but not nearly as bad as the gent fishing next to him whose T-Shirt bore the picture of a trout and the caption 'Take me to your leader'.) Anyway, the expert angler had an interesting method of retrieving his line. From a crouching position and with his rod tucked under his arm, he retrieved using a double-handed 'figure of eight' movement at a speed faster than a Frenchman untying his shoelaces in a brothel.

I needed to muscle up and be brave. It was only fishing, after all; nothing a few press-ups and limbering stretches couldn't solve.

I reached the swim and, as is my usual style, sat down and observed the water while I drank a cup of tea. There were fish moving all over the lake: big bow waves and bulging swirls were rocking the water and large dark shapes were cruising about deeper down. I concluded that, ignoring the size of fish and all the artificiality of a stocked lake, this was just fishing, still casting a fly to a trout and connecting with the hunter's instinct. As it turned out, it was fish on demand.

I'd been sitting down for three minutes, no more, when I became aware of activity near to the fishing lodge. The fishery manager had left the building, got into his pickup truck and was driving towards me. Had my payment bounced? Was there a rule that forbade bamboo rods? Were my wellies too muddy? Perhaps it was against the rules to bring one's own tea?

The truck trundled round the lake, eventually pulling up alongside me. The fishery manager wound down the window.

"I noticed you weren't fishing," he said. "Is sport no good?"

"Erm, sport's fine," I replied. "I was just taking it easy."

"So, you've not caught anything yet?"

"No."

"Not even a bite?"

"No, but…"

"Well we can't have you sitting around looking dejected."

Dejected? I was enjoying a lovely cup of tea!

The manager got out of his truck, grabbed a net from the back and ushered me towards some osiers at the rear of the swim. We pushed through the willows and, there before us, were three stock ponds boiling with fish.

"You don't want me to fish in there do you?" I asked.

"No, no." he replied. "I just wanted to show you my babies."

Babies? These fish were massive.

"How big?" I enquired.

"Smallest's only fourteen pounds," he said, shaking his head, "but there are some bigger fish in there to keep you happy."

I gasped, then made that long, blowing gesture usually reserved for expensive bills at car repair shops.

"Come on," said the manager, "get yourself ready." He then scooped up a net full of fish, carried them over to my swim, then dumped them into the lake with all the finesse of a publican emptying bottles into a recycling bin. The fish, bewildered by the experience, just milled about within ten feet of the bank.

"All yours!" said the manager, who then got back into his truck and sped off to the lodge.

What was I supposed to do? This was an unsporting 'ducks in a barrel' style of fishing. I looked down at those poor creatures. Theirs would be a short life. No catch and return. Just a short stretch of their fins before being hoicked back out and clonked on the head by a grinning punter. They were creatures destined for the gut-filled Dustbin of Shame.

I couldn't fish for them. It wasn't sporting and it

wasn't me. It would have been a transaction; money buying fish. There would have been no 'reading the water', no earning an apprenticeship, no 'studying to be quiet'. The pleasure of angling would have been bypassed in the rush for a pellet-reared Frankentrout so large as to dupe the angler into believing in his or her 'success'. And a management policy that frowns upon an angler sitting back and soaking up the atmosphere? What's all that about? One might as well return to work and order one's fish online.

I picked up my rod, net and creel and walked away from the swim. I heard a scurry of footsteps behind me and, within minutes, the screeching of reels.

Who was the real victor here? It certainly wasn't me. I was flattered by the generosity of the fishery manager and the hospitality of my employer, but having so much convenience laid out was like attending a banquet set for someone else. Sadly, the banquet – an obsession with big fish – is cholesterol blocking the heart of an otherwise healthy sport. I understand that these fisheries fill a particular need, but as an ex-girlfriend once said to me, "I don't care about your persuasions, you're not filling my 'niche'."

As with that girl, it was time to search of something better. A longer rod perhaps?

Stop – Unplug – Escape – Enjoy

Where do you go when you
dream about fly-fishing?

March

V

A NEW FLY ROD

Mrs H warned me that if ever I bought another fishing rod she would feed it into a garden shredder or, worse, feed in something else that 'wouldn't splinter'. She had good reason to complain. I'd bought several dozen fishing rods during a twelve-month period of 'bamboo indulgence' and then, after deciding that most of them were rubbish, spent seven weeks restoring and re-varnishing them so that they could be used.

"This is too much," said Mrs H as she walked, head in hands, into our bathroom that was filled with recently-varnished bamboo rods in varying stages of dryness.

"But it's the only room in the house that's dust-free," I replied, "and I only have another three rods to complete."

"One more week then, after which you will *absolutely definitely* need a bath, and a shave, and…"

"Another rod rack?"

"Enough! For nearly two months I have put up with having our bathroom covered in newspaper and fishing rods. The constant whiff of varnish is giving me a headache and you smell like the homeless love child of a rotten potato and a pickled onion."

"Seriously?"

"Yes. Your rods look great, but you look – and smell – a mess. No more. I beg of you."

"Okay, darling. I understand. Perhaps I should wash in the kitchen sink?"

"No!"

"Only joking."

"You'd better be. Besides, why do you want so many rods, anyway?"

After a long pause, and much thinking, I replied: "Because I *need* them."

Asking an angler why he or she needs so many fishing rods is like asking a woman why she has so many pairs of shoes. The urge to purchase such items, be they rods or shoes, goes beyond the desire to collect or hoard. It's as if something in our DNA tells us that without them we are somehow 'incomplete' – just empty hangers awaiting garments to give us purpose. The rational-minded person will say that one needs a range of items to accommodate all eventualities. A woman will want to accessorise differently for different events and an angler will want, if there is such a word, to 'tackleise' to accommodate every potential fishing condition. Of course, one's spouse doesn't always sympathise with an angler's need to own more rods that he or she has pairs of socks. And if a man has more rods than a woman has shoes, then there's no logical or emotional resolution. It's war. I realised this after tallying up the number of shoes in Mrs H's wardrobe. She has twenty-five pairs whereas I, after a large brandy, can declare that I have twenty-four split cane rods and a further fourteen made from other materials.

Thirty-eight rods isn't that many, is it? If I remember correctly, there are fifty-six species of fish present in UK freshwaters and it's only right, respectful and proper to have at least one rod for each. So I reckon I need another eighteen rods. But Mrs H wouldn't accept that argument. Let's just say that I'm in need of 'another' rod…

After my experiences last month, you'd be forgiven for thinking that I need to buy a twelve-foot ten-weight lion-taming whip of a rod. In fact, the opposite is true. I'm so against the mantra of 'size is everything' that from now on I will make a conscious effort to fish for smaller and wilder fish. To do so *will* require a new rod, and a new ruse to avoid the wrath of 'she who needs a bathroom'.

How, though, would I do it?

I decided to invent a story; one where I'd stumbled upon a shoal of burbot and, burbot being so incredibly rare, presented a once-in-a-lifetime opportunity to fish for them. I told Mrs H that burbot only fed in the third week in May and the only known method of catching them is on a Grey Wulff fished on a size 12 hook.

I also told her that my preliminary reading on the subject had explained that burbot, being such finicky eaters, would only succumb to an angler fishing with an eight-foot five-weight fly rod with translucent whippings and an olivewood reel seat, made by Edward Barder of Berkshire, England. ("Yes it's expensive darling, but I'm only thinking of you. You see, this one won't need any restoration...") I told her I didn't care if I only managed to catch trout on the rod; it was burbot I was after, and they wouldn't possibly take a size 12 Grey Wulff in the third week of May if I was using my normal eight-foot five-weight fly rod (the one with burgundy whippings and a walnut reel seat).

Mrs H had the power to grant me the fish (and rod) of a lifetime. All I needed was her blessing.

"Sweetheart, I *neeeeed* this rod," I said as I sank to my knees and pressed the palms of my hands together.

"Like you needed that pair of split cane carp rods," she replied, folding her arms, "that turned out to be so limp your friends nicknamed them The Brewer's Droops?"

"Eh?"

"The ones that cost you a thousand pounds but which, after you failed to sell them on FleaBay, had to flog at a car boot sale for a tenner."

"Oh, those."

"How is this rod different?"

"It's for burbot."

A NEW FLY ROD

"And where will you catch one of those?"
"In the river, in the third week of..."
"Sorry sunshine, I've heard it all before. No can do."
"Really?"
"Yes."
"Bugger."

Which brings me on to my request: I need to ask a favour of you. Given my propensity for out-and-out cowardice in the face of a scorned woman, I was wondering if you could test my next ruse to get the rod. You'll need the participation of your other half, but don't let her (I'm assuming it's a her) know that you're doing this for me. Play it straight and note her response. If it works, let me know and I'll try it. If it doesn't, then let me know anyway and I'll lock the shed door in case your other half asks Mrs H for a loan of the shredder.

Okay. Now remember, there's a new fly rod riding on this, so make it count. Here's what I'd like you to do:

Find a moment when your other half is looking especially relaxed, radiant, and receptive to romantic advances. (I'm told that the end of *Antiques Roadshow* is a good time.) Offer her a glass of wine and light some candles (women love candles, it's something to do with their fondness for moths). Then walk away – into an adjacent room. This is important. We don't want her to think you're after something. Act cool. Be natural. Put some Dire Straits on the music player and shuffle your feet from side to side as if you're using them to mix cement. (This is apparently very manly, making your feet subliminally assertive and, as I've found, warming up the soles of your shoes for when you need to make a quick getaway.) Walk back into the room. Tell her that you've been thinking about the good times you've had together and especially the time you and she went to *[insert details of event here]*. Say that you've been trying to think about what exactly it was on that special *[day/night]* that made you so proud to be *[married to/engaged to/going out with]* her. Then say that just then, when you were in the other room, you realised what it was. It was that she was so beautiful and that the dress she was wearing was perfect. She'll now either thank you or complain about your nauseating seduction technique. Either way, you should say that you're serious, that the dress has really got you thinking. And that, you have to confess, you've started to take an interest in dresses. (I hope, at this point, that she looks concerned rather than offering you free reign of her wardrobe.) I've read that a shock often puts women off their guards, so it's important to pause for a few seconds before saying that she shouldn't misunderstand what you're saying; that you don't want to *wear* the dresses, just that her fabulous taste in clothes has made you appreciate the importance of fashions and that it's important to buy quality things that make you feel good. Tell her she's lucky you're not a woman, that you're not competing for wardrobe space and that you don't have to change

A NEW FLY ROD

what you wear to match the seasons, and that you are, after all, just a hapless bloke who muddles along, asking only to be loved and for the 'occasional' opportunity to go fishing. Remind her that you have no 'little black dress', but if you did, it would probably be a fly rod that would be reserved for special occasions. It would be the one that made you feel happiest about yourself and the only one that would never go out of fashion. Tell her how you wished you had such a thing, so that when she looked back on her memory of when she'd seen you most happy, when she knew that you truly loved everything about your life together, that she would be imagining you with such a rod: perhaps, for random example, an eight-foot five-weight split cane fly rod with translucent whippings and an olivewood reel seat, made by Edward Barder of Berkshire, England. Ask her if she can visualise it. Ask her if she loves you. Ask her if she wants you to be truly happy. Ask her if she'll let you buy the rod...

Phew. Message delivered. Did it work?

Answers on a postcard to: Fennel Cowardly Hudson, c/o The Priory in Hiding, Bouvet Island, mid-Atlantic.

Stop – Unplug – Escape – Enjoy

What's your dream fishing rod?
When will you buy it?

April

VI

A VERY ENGLISH FLY-FISHER

Today is the 23rd of April: St George's Day, a day for Englishmen to hoist aloft St George's Cross (the England flag) and slay any dragons that cross their path. It's when afternoon tea is mandatory, wearing an out-of-season red rose in one's lapel is a must, and Elgar should be played on one's gramophone. It's a day to eat fish 'n' chips and smack the bottoms of our bottles of *HP Sauce* all over our English Breakfasts; when we should iron our trousers (not forgetting to warm our pants) while tipping our bowler hats in the direction of Her Majesty. It's when getting the bus to the pub, and drinking several pints of warm and yeasty ale, is compulsory. It's when *real men* will scoff jellied eels while *gentlemen* will be punting on the Thames. It's the day when Englishmen can dance the Morris Dance and proudly quote Shakespeare, shouting: "Cry God for Harry, England and St. George!"

You might be wondering what all this patriotism has to do with fly-fishing? Well, with the UK being the world's eighth-biggest tourist destination, and England being the spiritual home of fly-fishing, I'm feeling the need to brush up on my Halford and Skues. I'm aware that with the global accessibility of the Internet, there's a chance that you might be an *overseas reader*. Someone looking in, with expectations, a passport, and a travel rod.

Sadly, although I'm proficient at the Carshalton Dodge (a 19th century term for false casting), I am unable to direct you to the best beats on the Houghton Club's water on the Test, or invite you for a drink at The Flyfishers' Club of London. I can't even quote much Halford. The one quote I can remember merely undermines my credibility as an authoritative English fly-chucker. After all, did he not say, "No point can be of greater importance than a well-founded knowledge of where to cast; and it must be borne in mind that this knowledge is not intuitive, but must be acquired by marking, learning, and continuously studying the relation of the fish and its food, and striving as far as is practicable to take advantage of it"? So, I'm feeling the weight carried by English fly-fishers, who know that the world's fly-fishing heritage rests upon their shoulders.

It took me a while to become aware of the international appeal of fly-fishing. Sure, I knew that our patron saint of fishing, Captain Birdseye (famed for his impressive bread and frozen fish fingers), had navigated his way across the English Channel, where he is known as the not-very-English-sounding

Käpt'n Iglo in Germany*. But when it came to casting a fly, my vision was restricted to the 33-yard limit of my fly line.

How's that for international franchising? I guess it shows the global appeal of all things fishy. But heaven help the marketing man if he had to export my local butcher's 'Beefy Bob's Meaty Faggots Dripping in Salty Sauce'. Some things, it seems, are best kept to the land of Albion.

Recently an angling friend landed a job that required him to travel all over the world. One minute he'd be helping companies to improve their fortunes, the next he'd be fishing their local rivers. He sent me updates of his adventures via email. After a while, I concluded that if a country contained fish with adipose fins, or had ones with a taste for fur and feathers, then he'd be there. By very crafty design, he was able to work and fish in some (if not all) of the best overseas fly-fishing locations around the world. He experienced the Bahamas, Seychelles, Florida Keys, Montana, British Columbia, Alaska, New Zealand, Patagonia, Nova Scotia, Tanzania, Kola Peninsula, Mongolia, Lapland, Norway and Iceland. Not to mention the 'easily conquered' countries in between. Like France.

Obviously, I was hugely envious of his lifestyle and was thrilled at discovering species like arctic grayling, cutthroat trout, brook trout, steelhead, taimen, bonefish and permit. But was all the travelling necessary? All that gallivanting about when he had quality fishing at home? Did he *really* enjoy himself? Sure he did, the lucky sod. He caught more fish than a heron with a trawl net and, incidentally, met so many girls that he put more notches on his bedpost than a carpenter with a nervous twitch. But it had to come to an end. (I suppose there are only so many companies – and women – in Mongolia that require the services of a 'corporate services' specialist?) This week he returned home. I greeted him at the airport and received the shock of my life.

Colin (my friend) had left England as a ginger-haired and grey-suited middle manager. He returned as a bleached-blonde beach bum who insisted upon being called "Col" (pronounced "Cawl" – as though he only wanted me to speak to him via telephone). Gone was his English accent; his stiff upper lip had softened in transit. They were replaced by Bermuda shorts, reflective sunglasses and a strange little goatee beard that made me want to gnaw a hole in the skirting board. Colin had lost his English reserve. No longer was he the 'ever so uptight' Englishman who got sweats if he didn't eat roast beef on a Sunday; who ironed his socks and hung them in the wardrobe; who lived with his mum until he was 45; and who always brought cheese and pickle sandwiches to the office. He was now an international jet setter, who'd 'wetted his tackle' in virtually every dream location around the world. You can imagine the conversation that he and I had when we met at the airport:

"Colin, is that you?" I enquired.

"Yow! Fennel!" he replied. "Howz it hangin' Dawg?"

"So it *is* you then?"

"Sure it is, though it's *'Cawl'* now."

"Oh. So you won't be wanting these cheese and pickle sandwiches then?"

"No way Skin; quit 'em years ago."

"What about this Kendal Mint Cake? Or this thermos of milky tea?"

"Naah! Let's go sup a cold one. *Lowds t' tell ya abart.*"

Colin seemed to be as English as a bulldog wearing clogs, lederhosen and a beret. But after a pint of *London Pride*, he relaxed and started behaving like the Colin of old.

"D'you know," he said, "it's sooo good to be back in Blighty."

"Didn't think I'd hear you say that," I replied, "not after seeing your pictures of sea trout from Argentina."

"Or *Deeta the Dynamic* from Venezuela?"

"Cor, yeah. You told me about her. The things that girl could do with an ostrich feather, six Ping-Pong balls and a Chihuahua named Frank!"

"Don't get me wrong," said Colin, "I've experienced some incredible things, but there's something about 'England in Spring' that does it for me."

Colin's eyes had a glazed look, a dreaminess that told me he was either thinking of home, or picturing ping-pong balls appearing from unexpected places.

"Fennel," he said, "you and I grew up in a village that had only one road in and one road out of it. I needed to get out, to explore the world and see life. But now I'm back. And in returning home, I realise that I've been away for too long. All my travelling, all my dreaming, has led me to one conclusion: that I want to be here, fishing my local rivers, where I am as much a part of the landscape as it is of me. I want to sit beside the river, drinking Kelly Kettle tea and listening to the songs of woodpigeons and blackbirds. I want to hear church bells ringing on Sunday. I want to look up and see May blossom in the hawthorns. I want to smell meadowsweet and honeysuckle at dusk. I want to savour the English weather, get wet when it rains and thank the sun when it shines. I want to wear tweeds and brogues, and don a *very English* hat. I want to see a trout rise and cast my Greenwell's Glory towards it. I might catch the fish, I might not. It won't matter. There's a 'knowing' in such things, a 'being' in such places. It's where we're meant to be. Our ancestors knew this place, and our children will too. And so long as we're here, it will be ours, too: in our hearts and dreams, and on our flag, calling us home, to live and be free. This time. This life. This *England*. It's what it means to be an English fly-fisher. It's why I'm glad to be here, to be *home*."

Colin and I looked at each other and smiled, knowing that nothing more needed to be said.

May

VII

IF ONLY EVERY DAY COULD BE LIKE THIS

Colin was right: fly-fishing in England has a sense of homeliness to it. Inspired by his words, I have decided to have the most English of fishing days, one where dreams and reality blur to create an image of the ideal. The plan is as follows:

This morning I shall rise, dress myself in Derby tweed – breeks, waistcoat and a Norfolk jacket – then descend the stairs of my dreamed-of cottage to make myself a cup of tea. I shall open a tin of Ceylon Orange Pekoe, smell its sweet aroma, then sprinkle the leaves into a teapot while the kettle boils upon the Aga I yearn to have. Breakfast will be freshly-baked granary bread (two slices, toasted, with butter from the pantry) followed by a deep breath of morning air as I step into the garden.

Later I shall visit my study, selecting a bamboo fishing rod from the rack and a creel from behind the door. I'll venture into the hallway and select a hat that suits my mood; thinking, "Today is a Trilby day – too warm for a deerstalker, not windy enough for a cap and not sunny enough for a Panama." But when the time comes, I may prove fickle and select my favourite pudding cap. I shall have to wait and see.

I will fetch my bicycle from the outhouse, check that its chain still squeaks enough to annoy Mrs Dunnock at Number 44, then strap my fishing rod to its crossbar, shoulder my creel and cycle forth along the lanes until I reach a crossroad. Left will be the river; ahead will be The Swan. A pleasurable dilemma: should I fish or head to the pub? I imagine it will be both. Ale first, river second.

Later I shall explain to the trout that a ploughman's and a pint of beer was a better way to spend my lunchtime than chasing their rises. They will understand my decision and thank me for the respite.

When I eventually fish, it will be in the slower water away from the weir. I might rise a trout, I might not; it doesn't matter. Maybe I'll abandon all thoughts of fishing and opt for an afternoon nap in my usual place beneath the willow? Yes, I like the thought of that; I'll make time for the snooze, whether I fish or not, and leave my pocket watch unwound.

The day will end much as it began, back at home, with a cup of tea in hand and a tour of the garden in the half-light. I might opt for an early night and the comfort of warm blankets, perhaps with a favourite book for company, or I might stay up late to write by candlelight.

Maybe I won't do any of that, instead opting to stay in bed all day?

Whatever I decide, today will be a fine day of fine things; a day spent on my terms. For I am a traditionalist; a gentleman angler. Proud to be eccentric; delighted to be different. I seek contentment, catching the mood on the breeze and sensing the current of life that pulses through the moments of calm that exist in the backwaters of our busy lives.

I am free, so I will go forth and fish – as a man alive.

Stop – Unplug – Escape – Enjoy

Describe your perfect fishing day.

MAY

VIII

THE AMWELL MAGNA FISHERY

Let's say you've recently won the Lottery. You've recovered from the shock, had your holiday of a lifetime, bought your dream home, treated yourself to an automatic Teasmade (which sits, pride of place, on your bedside table) and have invested in a small company that makes authentic silk fly lines. What then do you do? As an angler you'd probably start thinking of buying your own lake or stretch of river, somewhere you could erect a fifty-foot sign saying "Private, Keep Out!" (Or, more accurately, "Millionaire in Residence, Clear Off!")

Having a water all to yourself, and the money to turn it into your perfect fishery, is the ultimate dream for an angler. With enough money, you could start from scratch and dig a custom-made pool or restore a substantial length of river. But if you *were* a Lottery winner with enough money to date a supermodel and get her to pick up the dinner bill, then you'd want to cherry-pick the best and most historic water you could find. You'd approach the owner, saying, "Here's a cheque. You fill in the numbers. Don't bother finishing your breakfast; I'll eat that. Just leave the keys on your way out." A trout fisher might purchase the Houghton Club's water on the Test, a salmon fisher might purchase the Glendelvine beat of the Tay, a grayling fisher might purchase the Corwen stretch of the River Dee. If money were no object, where would you purchase?

Unfortunately, most of us are not Lottery winners. We don't have Elle Macpherson's phone number and we have to get out of bed to make a cup of tea. We have mortgages, car loans, overdrafts, and barely enough money left at the end of the year to buy another fishing rod. Having our own lake or river is little more than wishful thinking. We have to 'make do' with a mixture of day- and season-ticket fisheries, club waters and sneaky visits to 'off limits' places. But what if you were presented with the opportunity to lease or purchase a perfect river or lake? Maybe it's one you've fished before and know of its history and potential? You'd need to act quickly before someone else snaps it up or you found yourself in a bidding war for water that's doing its best to slip through your fingers. But you haven't the Lottery winnings to shower upon the opportunity and your 'three shillings and sixpence' life savings isn't going to secure the winning bid at an auction. What do you do? You put the word out to like-minded friends so that, together, you can make an acceptable offer. This is how many syndicates are formed and how, back in the 1830s, the UK's oldest angling club (that still fishes the same water) came into being.

The Amwell Magna Fishery, just north of London, provides one-and-a-half miles of trout fishing on the 'old river' Lea and its tributary the River Ash. It is where Izaak Walton, author of *The Compleat Angler*, fished in the 17th century and where I like to think he wrote the lines: "I care not, I, to fish in seas, fresh rivers best my mind do please," and "I have found it to be a real truth, that the very sitting by the river's side is not only the quietest and fittest place for contemplation, but will invite an angler to it." Which explains why I'm so excited at present. You see, I am writing this sitting beside a weir pool on the Amwell Magna Fishery, as I have done each year for the past decade. I am in 'quiet and contemplative mood', keen to reflect upon the importance of this fishery both as a historic water and as an example of how anglers in their association can achieve great things and safeguard the future of something special. Let me begin with the past, and then work my way up to the present, sharing the story of the Amwell Magna.

Cast your mind back, in soft rosy focus, to the early 19th century. Sir Francis Chantrey and his friends William Whitbread and Henry Warburton are founding the Amwell Magna Fishery. Each is from London's High Society and all are members of the Houghton Club. They take a lease on the water in 1837 and then, in 1841, pass the lease to William Shackell (creator of the ink used to print Penny Black stamps) who runs it as a subscription water with each angler a 'Member of the Amwell Magna Fishery'. Henry Wix arrives as Club Secretary in 1851, attracting prestigious members including Knights of the Realm, an Admiral, a Major, a Lieutenant Colonel, Members of Parliament, doctors, lawyers and merchant businessmen. R.B. Marston, editor of the Fishing Gazette, is associated with the Fishery from 1894 and then, in the second half of the 20th century, the actor Michael Hordern becomes a member. That brings us up to today, with the fishery having a discrete membership including music and television celebrities, a Knight of the Realm, professional gentlemen and retired individuals.

The most impressive thing about the Amwell Magna Fishery, however, is not its list of members but the way in which the river has been cared for during these past 173 years. Club records show that 'considerable work' was done to the river in the 1830s to turn it into a fishery and that full-time keepers were employed for over a hundred years. (In the early 20th century, the Fishery employed two keepers and fourteen river watchers.) The club began a trout breeding and stocking programme in 1855 and after the Second World War planted half-a-million mayfly ova into the river. In recent years the club has undertaken extensive revetment work, narrowing channels to improve flow and clear sediment.

And last year they removed 14,000 signal crayfish from the river (averaging 275 crayfish every three days during the warmer months). The members, with support from the Wild Trout Trust, have also thinned overhanging trees from their stretch of the River Ash, enabling greater light penetration and weed growth. The result of these recent actions is a noticeable increase in spawning redds and wild brown trout in the river, an excellent mayfly hatch, and classic dry fly sport. But above all, the members have demonstrated their commitment to upholding the traditions of river fly-fishing for trout. (The fishery is a strictly 'upstream dry and nymph' water where members fish light tippets with small flies that 'match the hatch' through the season.)

As Bernard Venables wrote in his book *Fishing, British Sports Past and Present*: "As the numbers of anglers increase and, proportionately, waters decrease, the future must inevitably be with clubs more than with individuals." At Amwell Magna, the future is in great hands. The fishery benefits from its long history, but it's what the members are doing today that makes it such an exciting place to fish.

Members, custodians, guardians. Call them what you will. Together they are keeping alive the legacies of Izaak Walton, Francis Chantrey and Henry Wix; and they're perpetuating the tradition of trout fishing in the Lea. Without them, the trout would have long-since departed and the future of this lovely fishery would be very different.

Which brings us up to the present. As I was saying, I am sitting beside a weir. The river above it is slow and the trout are rising lazily to mayflies. Below the weir, the water is tumbling and swift, yet I can see mayflies rising at its tail. Here the river splits into a millstream and narrow river. Neither are more than twelve feet wide, making them ideal for stalking trout that lie among the lush weed fronds and beneath the trees. I haven't started fishing yet – that will happen after lunch – as I've spent the morning walking the river and exploring the water meadows (which are alive with mayflies and early damselflies). A hobby is darting and gliding above the cow parsley, catching its breakfast. Bitterns are known to frequent this area (there are lakes adjacent to the river) but the highlight for me is a pair of Cetti's warblers that are nesting opposite the clubhouse. I've never seen or heard these birds before, and have been captivated by their astonishingly loud bursts of song and accusations that I have a "cheeky leaky seat".

I'm here as a guest of Huw Williams, Chairman of the Amwell Magna Fishery. Like me, he loves the river and its history. A jovial and welcoming host, he also has a quiet side that tells me he knows both the value and fragility of this place. While many traditionalists harp-on about the 'glory years' of angling, when the countryside was quieter, when rivers had more water and a man could fish merely for the pleasure of 'being', my host knows that the glory years are today and the future, given that they're ours for the making. He says, "If we're not content with what we've got then we have to go out there and make it better." Wise words for anyone in control of their destiny and those who can influence their environment.

Huw's leadership, and the continuation of quality fishing at the Amwell, is testament to his beliefs. It proves what can be achieved when the right people come together to care for and nurture something they love. Their collective energy and vision, this 'society' of anglers, brings strength to their being. So much coming together, so much growth and evolution; it makes me proud to be here, to value traditions and appreciate the quality of time spent contemplating one's good fortune.

As I sit quietly beside the river, watching its water move purposefully downriver, I realise that time is constant, perpetual and relentless. It moves no quicker or slower today than it has ever done, or will ever do. Yet sometimes it feels like it's as fleeting as dust on the breeze. At places like the Amwell, one can hold out one's arms and feel time collecting in our hands. The weight we feel in our palms is the responsibility of heritage, keeping the past alive but in doing so, creating a new future.

If I had all the money in the world, and could fish wherever I wanted, where would I cast my line?

On a day like today, when hobbies glide across a deep blue sky and mayflies rise to the sun, I would be here, as I am now, at The Amwell Magna Fishery, sitting contentedly, contemplating the richness of an angler's life, while studying to be quiet.

Amwell Magna is a fishery that knows its place in time. Here the future is informed by the past and influenced by the present. It gives us faith in the beauty of the world and man's ability to care for it. In the words of Izaak Walton, it's somewhere to "look about you, and see how pleasantly that meadow looks; nay, and the earth smells so sweetly too." Fly-fishing takes us to such places. The 'tick-tock' of our casting rhythms becomes more of a 'tock-tick' as we wind back the clock and savour what was, is, and will be again.

May

IX

LOCAL FISHING

May is here and with it a new mayfly season, when anglers embark to favourite waters, many of which are far from home, in search of up-winged and adipose-finned dreams. I have a brief moment to indulge this vision, albeit one that's not so far away, so if you'll allow me then I'd like to share a mini fishing adventure with you. It's written in note form, directly from my diary, to give a sense of immediacy to the adventure.

I returned home from work today at 6.30pm. I quickly went inside, collected my fishing tackle, and then walked briskly towards the river, slowing my footsteps and removing my cap as I passed a war memorial on the village green. Cherry blossom carpeted the ground around the names of fallen men. I paused to reflect and honour them. Then, after the quiet interlude, I made my way across a water meadow to the river. There it was, looking grey and un-spring-like. The slow-moving water reflected the bare branches of alders and willows above. But – glory to behold – the meadows were laid golden by millions of celandines. Jackdaws hopped through the grass, searching for worms, and long-tailed tits chattered amongst the hedgerows. I smiled when hearing the frail but enthusiastic bleating of week-old lambs, born late but grateful given the recent harsh winter, and cringed when I heard the sinister cackling of magpies leering over them.

I continued walking, feeling my heart lift as I arrived at my chosen spot on the river. The greyness was gone. The water was clear, the gravels were bright and the weed almost luminescent. Like a cathedral window at sunrise, the river shone with brilliant greens and golds. And there, below some stepping-stones in the river, was a shoal of fingerling trout cavorting like a murmuration of starlings as they fed on their mid-water supper.

And then I heard a splash. My gaze became sharper in its focus. What had made the commotion? A kingfisher? A mink? Or a trout? I turned around and scanned the river for the source of the sound. There, thirty yards upriver, was the ring from a rising fish. I watched closely and saw another rise-form in the same spot. A good-sized trout was feeding confidently, its pectorals breaking the surface as it rolled. I walked slowly forward, then stooped down, creeping towards the trout. Soon I could see what the fish was taking: olive duns attempting to flutter free from the water. The trout was a wild brownie, twelve inches in length, looking like it had earned its claim to the best spot in the river. Here was a challenge to test the fumbled casting of this early-season angler.

A look through my fly box resulted in raised eyebrows,

an open mouth, and a subtle curse. I'd brought the wrong box – the one still in my pocket since my late season reservoir fishing. As I flicked through the sedge and daddy-long-legs patterns, I wished for a Greenwell's Glory or a 'parachute anything'. Nearly all the flies were too large and too clumsily tied to pass as the size 16 olives fluttering above the water. In the end I plumped (quite literally) for a size 14 elk hair caddis. At least it was brown and likely to float.

I knelt beside the river, just ten yards downstream of the trout, and made my first cast. The rod felt vibrantly alive as the bamboo flexed and the line wisped rhythmically out across the river – straight into the branches of a coppiced alder. I paused, while 'releasing' the appropriate expletive, and then tugged at the line. The fly sprang free. I drew in the line and cast again. The fly landed just upstream of the trout and, as it drifted past the fish, the inevitable happened: a calm and slow-motion display of *absolutely nothing*. I could almost hear the trout laughing as the straggly piece of moose bum floated over its head. Six successive casts resulted in the line catching most things that moved, and many things that didn't. But I couldn't hook the trout, which continued to feed exclusively on the olives. I felt my teeth clench and my blood pressure rise. And then I laughed.

With all the bother of attempting to catch a fish within such a short window of opportunity,

I'd blind forgotten to check the time. I reached into my pocket and noted the hands of my fob watch. They showed 6.50pm. I still had another half hour of fishing ahead of me. It was time to slow down, relax, and *breathe*.

There's a knack to this 'fishing slowly', where doing less is doing more. I reeled in and crouched beside the river, just watching the trout feeding. And while I did, the sky lightened and the sun shone through a gap in the clouds, warming the landscape with its coppery rays. The breeze dropped and the river became almost mirror-calm. I listened to the onset of a near-silent dusk, where the splashes of trout and the flapping of pigeons going to roost were the only sounds keeping the day awake. Soon they were met by the clicking of a ratchet as I pulled another five feet of line from the reel. *One last cast before I return home.*

You can guess what came next: a confident take, a trout jumping on the end of a bow-taut line, a jagging fight and a victoriously-clenched fist as the trout was scooped into the net. I brought it ashore, laid it gently in the grass, photographed it, and then returned it to the river. I had caught the fish that would feature on the cover of my fly-fishing book.

The whole finale took less than two minutes but seemed to last for my allotted hour. For this was all I had: an hour at the end of day. It was all I needed – to connect with the river and feel like I was properly home. But I shouldn't have needed to catch that fish. It was, after all, my neighbour. It would still be there tomorrow.

Roderick Haig-Brown wrote in his *Fisherman's Spring*, "There will be days when the fishing is better than one's most optimistic forecast, others when it is far worse. Either is a gain over just staying home." But if the river and one's locality are all part of 'home', then I'd rather stay put and savour familiar surroundings. This, after all, is where one's heart resides. We create it. It is ours to define and enjoy. My river, that first trout, and this Cotswold valley, are as much a part of my home as the walls, windows and tiles of my house. Being here is about looking for the angle of light that forms a rainbow, and searching therein for one's dreams. If you visit, don't walk too far. Beauty is already within range. Shorten your line, focus your casts, and slow things down.

Enjoy the magic of local fishing.

Stop – Unplug – Escape – Enjoy

What's your favourite local water and why?

May

X

FISHING, FAR AWAY

Do you remember me mentioning The Flyfishers' Club of London, the esteemed gentleman's club whose patron is Prince Charles and whose membership includes pretty much every notable English fly-fisher in history? It's the one I described with an 'if only' remark when thinking about visiting its hallowed rooms in Mayfair. Well, it turns out that I was invited to join the establishment in recognition of my articles about fly-fishing. As part of the process, Peter Lapsley, Editor of the Club's prestigious publication *The Flyfishers' Journal*, asked me to write an article for him. Alas, he died before I had chance to submit it.

Theo Pike, the Journal's new Editor, took up the reins of his predecessor and nudged me to complete the piece. The article, about local fishing (which you have just read), was published and I gained membership of the exclusive club. I was delighted on both counts, and inflated my ego by telling everyone about my newfound status. "There's more to fishing than catching fish," I boasted, quoting the Club's motto *piscator non solum piscatur*, "especially when it's on one's doorstep or involves lunch with good friends."

Alas, I didn't know the half of it.

Tim Pike, my friend and loyal reader, had sponsored my application and was thrilled that I was accepted to the club. Soon I would be joining him as a fellow 'young member'.

"Just wait," said Tim. "Your angling world and social circle will change forever. Those local waters of yours will flow into distant rivers and oceans; your acquaintances will broaden and your fishing opportunities will be limitless. I predict that the quality of your life will be *infinitely enriched* by your membership of The Flyfishers' Club."

Crikey-o! What a prospect for a chap who'd gone into print saying that local fishing's the thing to do! Best I grab a train ticket and head into town...

Sure enough, within a week of joining the club, Morgan Jones (who organises events for the young members) sent me an email. A club member was offering the chance to join his party for a week's salmon fishing on the Moy in Ireland. Everything was paid for, save the cost of travel. All I'd need to do was jump in the car, board a ferry, and wet a line. What an intensely exciting (and slightly terrifying) opportunity. It was almost too good to be true. But it was very real. I was staring straight into Santa's goody bag.

Did I go? I did not. It was all too fabulous, all too soon. I cited seasickness, an allergy to Guinness and a phobia of leprechauns as the reasons why I couldn't attend. I'd wimped out when secretly I would have donned a leotard and taken the lead role in *Riverdance* to go there. I picked myself up off the floor and carried on reading my subscription copy of *Small World Monthly*, half hoping that the invitation would not be my last.

Mid-July came and with it the blazing sun and blue skies of a proper summer. The trout in my local river began sulking and I found myself sitting beneath a cherry tree in my garden, sipping a glass of iced gin & tonic. H.R. Jukes' *Loved River* was on my lap and – potential disruption that it was – my mobile phone was in my pocket. I was preparing for an afternoon's reading when the phone dinged twice, signifying the arrival of a text message. I checked it and, to my delight, received news from Tim that he was on the riverbank. It said: "Reading the river. Taking my time. Excited to be here. Big fish likely."

A member of The Wilton Club on the Wylye, Tim had recently been 'properly smashed up' by a twenty-inch brown trout that had taken a size 16 dun. Maybe he was after the fish again? Surely not in the heat of the afternoon sun?

Another text arrived, again from Tim. It said, "Wading easy. River broad. About to cast in. Somewhat nervous. Fish can leap five feet from the water when hooked."

Five feet? On the Wylye? Surely they'd be taking

off and landing in Salisbury?

I replied: "Gin was strong in the clubhouse was it? You'll be telling me you've caught a double-figure brown next!"

A minute passed before the phone chimed again. Another text from Tim, this time shorter than the others. Just two words of infinite intrigue, saying: "Well, actually..."

And so the exchange continued, with me becoming ever-more confused and intrigued by what Tim was doing. Texts arrived saying things like, "Beautiful sunsets," "Icy cold water," "Massive, dramatic waterfalls," "Home of the gods." Hmm. Where on earth was he fishing?

The book in my lap gradually lost its appeal and the ice in my glass melted. My local river ceased flowing through my thoughts and a hoped-for ten-inch specimen withered to a three-inch cold-water parr. Suddenly I had nothing to brag about. I crossed my legs, put down my phone, and awaited further news. It was then that photos started to arrive: images of Tim fishing a river wide enough to be a highland loch; a waterfall that made Niagara look like the dribble from a spilled pint glass; one showed Tim's rod bent double and was accompanied by the words 'Land of Giants'; and finally, a picture of him holding an intensely spotted large-headed fish that looked more like a sea trout than a brownie.

FISHING, FAR AWAY

I replied saying, "C'mon, tell me where you are!" then received two more messages: one an image of a fish held underwater, which looked like an Arctic char; the second of a flag which, if my geography lessons at school were any good, was that of Norway, Iceland, or Aston Villa football club. Sadly, it would be another week before I knew the answer.

I'd all but forgotten about the exchange of messages when a letter arrived from Tim telling me of his recent adventure. It transpired that he'd been fishing from Raudholar Lodge on the Upper Laxa in Iceland. The trip had been arranged through Flyfishers' Club member Peter McLeod of Aardvark McLeod and was hosted by Charles and Alex Jardine. Tim didn't really know what to expect, other than he'd been briefed that the minimum recommended tippet is 5lb and that the local liquor called Opal is the cause of many a missed flight at Reykjavik airport. His first impressions of the river, at Laxardal, were that it was huge: "the width of a football pitch" and set in outstanding scenery: "think west coast of Scotland, but more dramatic and with spectacular lava fields." He began fishing on the afternoon of his arrival, being guided by Guðmundur (Gummi) Bjarnason. This young man could smoke a packet of cigarettes in less time than it took Tim to lose his first fly (something that happens quickly with Tim). Tim felt slightly out of his depth fishing this vast and new river, but soon realised that the principles of fly-fishing are the same wherever you go – it's just that anglers have their quirky local preferences. Dry fly was the exclusive technique and the fish – which were unlikely to have seen a hook before – fought savagely. They averaged in excess of five pounds, with Tim's largest specimen weighing 9lb. With the benefit of 24-hour daylight, fishing days were long. First casts were made at 8am, and last ones at 10pm, with much hospitality in the lodge after this. But the most bizarre thing Tim mentioned was that it was possible when wading to have feet numb with cold one minute and then, after a quick shift to the right, to feel as if he were standing in a hot bath. Having never experienced this before, it made me realise the extreme differences in what is familiar and local, and what is unfamiliar and distant. Each can be an adventure, but this far-off fishing presented such alluring challenges. There would be none of my usual 'grab the rod from the porch and stroll down to the river' approach. Instead, it would involve travel cases, passports, time zones, exchange rates, airport lounges, super-strength line and the utterly terrible prospect of having to buy *even more* tackle.

I like the thought of fishing overseas, where everything's bigger, pulls harder, and is there for the taking. But I fear that I'm too much of an amateur for fishing like that. In fact, I'm a right duffer.

May

XI

A DUFFER IN MAY

Learning to Love

Being a writer is very similar to being a fly fisherman, insomuch as we put something out there in the hope that it will look appealing. If our timing is right, we might 'make contact' and bag some friends and memories from our well-placed lines. Mostly it involves more time writing the words than people spend reading them, but it's worth the effort in the hope that one or two special connections will be made. A reader *who understands* what I'm talking about; who says, "I like the look of this book and this author," and takes interest. Just like Tim Pike did when he read my work and then sponsored my application to become a member of the Flyfishers' Club. It's the most rewarding and cherished outcome from all of my writing activity to date.

A writer-angler is grateful for his or her red-letter days; but success is never *expected*. If it were, there would never be special red ink in our journals or diaries, only the normal colour (which, in my case, is a burgundy-brown). Our hearts wouldn't race when a book is sold, or flutter when a trout rises to our fly. Life would be terribly dull. So we busy ourselves by enjoying the process as much as the result. It's not about the number of books we publish, or fish captured; it's about doing what we love.

Fishing, as I see and practice, is creative *input*; whereas writing is creative *output*. One feeds the other. Experiences give me stories about which to write (and, to a degree, my writer's eye enables me to spot 'angles' in the things I experience). The lesson? That we should respect angling as an *experience*, rejoicing in the simple pleasures that it brings. And then, when we've calmed down, we should write about it so to never lose the memory.

Angling is like the lover who generously and passionately gave us our first kiss, who stole our heart and set the unforgettable benchmark for all who followed. 'The one', the first, the original. Yet, inevitably, we crave more or 'different'. With appetite whetted, we seek the riches of what might be 'out there'. So we practise our chat-up lines, shine our shoes, comb our hair, and see what we can pull from the stew pond of life's possibilities.

We venture forth, in hope.

Do we capture what we seek, or do we find our eyes being caught by passing interests? Angling always leaves us hungry for more. But eventually we find ourselves drifting back to the simple pleasures of the art, retracing our footsteps to waters or fishes of our youth, to rediscover the 'missed heartbeats' of our first kiss. It's there where the discovery was strongest.

Always remember the blend of terror and excitement in hooking and landing one's first fish. Being humble and grateful help us to appreciate life. Thus with angling, and fly-fishing especially (as it's the simplest and purest form of the sport), it doesn't pay to measure one's happiness by the weight of one's creel. There's rhythm (in one's casts) before the climax. So take your time. Enjoy it.

Never, ever, stop to question whether your performance is the best.

Get Some Action

No form of angling should be competitive, at least as I see it. It is, after all, just a recreation. It shouldn't matter if we catch less or more fish than the next angler. Sure, it's nice to see *Hesgot Atgani** arrive at the river, full of arrogant swagger, only to storm off in a huff when he fails to rise a fish on his first cast. But he's a lost cause anyway and probably will be swapping his rods for golf clubs before you've finished your sandwiches and reached for your fly box. You're different. You've taken your time. You know how to 'Stop – Unplug – Escape – Enjoy'.

*Hesgot Atgani is the name I give to an angler who

attempts to bypass his angling apprenticeship by kitting himself out with all the latest tackle before he's learned the rudiments of the sport. It stands for 'He's got All The Gear And No Idea'. The female equivalent is 'Shesgot Atgani'.

While I'll always harp on about angling's magic having little to do with the act of catching fish, I have presently found myself in an unusual position. I have a deep sordid desire to get my rod into action. Like a sailor who arrives in port, I want to get out there, have some fun, and *catch something*.

I've done a lot of talking this year, but not a great deal of fishing. I've caught only two trout so far in this book: one in my photo album and one from my local river. I am, as you've probably observed, enduring a relatively fishless season. Which is not good. You'll soon be calling me *Hesgot Atwant* (He's got All The Words And No Trout). So I ought to offer you a glass of Flyfishers' port and see if you'd like to drink.

I'm going to set myself a challenge. I shall go to the Windrush, at the time of year when trout are rising most confidently (thanking the mayflies for helping to ease the challenge); I'll stand proudly, like a peacock beside the river, and muscle up to whichever angler seems to be 'doing the business'. I'll then display my trademark 'triple-whip-spiral-zip-swallow-loop-wiggle-wrist roll-cast' technique, and have every trout in the river begging to be next onto my hook. I'll get cloth badges embroidered with the word 'Troutmeister' and sew them onto my fishing waistcoat. I might even get some of those 'wrap-around' sunglasses that provide heron-like vision and allow us to move in slow motion, just like in *The Matrix*.

But hey, it's only fishing. I can take it in my stride. After all, the only things I need to assist me are a suitably 'proud' hat and a box full of mayfly patterns. And perhaps some cake for elevenses. And maybe a rod and some line. But enough of the detail. Let's get keen, get eager, and get fishing!

Over Too Soon?

Two hours have passed. I'm now standing beside the river. Swallows are swooping and chattering on the breeze, the sun is high and the mayflies are up. I have my fishing rod in one hand and 'something else' in the other. I'd like to sit down but I can't. I should be wearing trousers but I'm not. I'd like to show you some fish. But I can't. You see, things haven't exactly gone to plan. In port terms, the glasses are all over the floor. Let me explain, but before I do, I should tell you that I'm somewhat short of breath and nursing an uncomfortable throbbing 'down below'.

The day started well. I arrived at the river to see its waters running steady and clear, with trout holding in the expected positions in the flow. There were three other anglers present, one of whom was wearing an especially 'uncamouflaged' stone-white waistcoat. (It would appear that he'd just returned from a fishing trip to the moon.) He was casting with crisp but fluid movements, sending out loops of line that were so tight that they travelled through the air shouting "Humbug!" This angler was to be my benchmark, the one I'd challenge to prove that one doesn't need lessons or modern tackle to catch something special.

I assembled my 7ft 6in Garrison-taper split cane fly rod, threaded it with a 4-weight double-taper fly line, attached eight-feet of leader, tied-on a size 10 Walker Mayfly, then shouldered my creel and walked with bold steps towards my opponent. I stopped twenty yards from him, observing his lightning-fast reactions, efficient striking, and mechanical playing of fish. (I named him *Courtney Fish*, as this seemed to be his only interest.) He was good. Very much 'in the zone'. A proficient hunter, ready for a beating.

I continued walking until I was fifty yards upstream of Courtney. Far enough away as to not spook his fish but close enough for him to see me in action. And whilst he was wading, I opted to fish from the bank, standing in proud silhouette against the sun. I didn't worry that I might put the fish down; I wanted to dazzle my enemy with my brilliance, blind him with flashes of sunlight upon my impeccably varnished rod, and humiliate him with my inherent fish-catching ability.

I studied the water. Three trout were rising within

casting range. One of them was leaping clear from the water to catch mayflies in mid-air; the other two were rising so savagely that they could have been eating ducklings let alone mayflies. And they were big – at least twenty inches apiece. The ultimate opportunity to outdo Courtney in size *and* number of trout. *This would be easy.* I rubbed my hands, knowing that I'd have to be a complete idiot not to catch.

It's Only a Little Prick

The Collins *Concise English Dictionary* describes 'idiot' as "a person with severe mental retardation; a foolish or senseless person." Hmm. Harsh words. Perhaps, if I asked the lexicographer nicely, he or she might soften the description to "Blond-haired Englishman, known for wearing tweeds and writing journals; has fine taste in fountain pens and fishing rods and has no concerns about being unable to catch a trout on a day when a blind comatose sand lizard could hook a fish."

Under normal circumstances (when nobody's looking) I can catch a trout quite quickly. But not today. Today I didn't even get to wet my line. Just as I was about to cast, I glanced over my shoulder at Courtney. He'd stopped fishing, was standing in the river with his rod vertical, and was displaying a stare that said, *"Go on then, show me"*. It was now or never. I pushed out my chest, limbered up my arms, tilted my head –

in a sort of John Wayne way of saying: *"I don't care if them's Injuns, I got a loaded gun!"* – and readied myself to cast. I lifted my rod, pulled some line from the reel, and began to false cast, letting out more and more line with each forward movement. I got about ten yards of line in the air before doing some fancy snake-like swirls (just to look good) and then the rod reached the limit of its power. I wasn't going to let this stop me, so I tensed my bicep and lifted my arm higher, hauling the line back through the air. The fly hurtled towards me, across the water and up and over the bank, then did a last-minute twist before thumping sharply and deeply into something 'personal'. I doubled over, dropped the rod, and let out a scream. I fell to my knees, grabbing my crotch as I did so. I looked down. There was the fly, hooked firmly into the nether regions of my trousers. Its point and barb were lodged into something that wouldn't normally rise to a mayfly.

It isn't easy to stay calm when you can barely see for tears, and especially while someone is watching your every move, knowing that you have a fishing hook buried deep and piercingly in the vicinity of your 'unseeing eye'. But pain sure focuses one's mind. I found myself thinking: "Is it a barbless hook? Do I have any forceps? Should I ask for help? Will I need to go to hospital? (And if I do, how on earth do I explain what's happened?)" And, rather important to an English gentleman, "Am I wearing clean underwear?"

And then the realisation: I had answered 'no' to all five questions. *Bugger. I would have to deal with it myself.*

Trying to remove one's trousers and underpants, while having one's manhood pinned to the garments, requires Houdini-like flexibility and a breathing technique akin to that practiced in childbirth. It took a lot of painful tugging to get my first leg free, with me lying on my back and my legs gyrating in the air, and I wondered whether it would be easier to cut the garments free with a knife. Eventually, after much squirming, I removed them and stood up. My undies and trousers were hanging from my Hampton like flotsam on a branch (okay, 'on a twig'). And then I could see what I was doing.

There is was: my poor, sorry, todger. There was the fly, lodged in the unexpected 'member' of the fishing club. And there was the blood, soaking my clothes and running down my leg.

Jeez. I would have to apply pressure to the wound *and* operate. Two hands: one to squeeze the circulation from my pecker and one to gently, calmly, and painlessly remove the hook. Well, that was the plan.

Sweat began pouring down my face and the world began to spin. I reached down to the 'scene of the crime' and, with eyes closed and my stomach churning, attempted to wiggle the hook free.

It wouldn't move.

I applied some more pressure.

Nothing.

I tugged at the hook.

Bad idea.

I opened my eyes and saw a blood-red mayfly snarling back at me.

And then the worst thing possible: I felt a hand on my shoulder and heard a voice saying, "Are you okay?" It was said with the condescending sympathy of someone consoling a puppy that had been left outside in the cold.

I turned and saw Courtney peering down at my trouser snake, his head leaning away from me as if to say, "I need to look, but I know I'll regret it."

"No, really," he continued, "are you okay? That looks, erm, nasty."

"Not exactly," I replied. "I got 'caught up' in the moment."

"I can see that," he said, "fancy you trying to outperform me with that little bit of wood of yours? It barely passes as a rod, y'know."

"Pardon?"

"The fishing rod."

"Oh, okay, I understand."

Courtney knelt down beside me and took a good look at the offending fly.

"Whoa, mate! That really *does* look nasty!" he said as he took a sharp intake of breath. "Is it shop-bought?"

"What?!" I exclaimed.

"The fly, it's at least three sizes too big. But it seems that it's capable of attracting a tiddler."

"Very funny."

"Lot of blood there, though. Could get infected, become gangrenous even. Might need surgery. Possibly an amputation, if it doesn't drop off on its own. Yep, it's going to be 'needle, scissors, and a bandage' for sure. Or possibly a splint, if they can find one small enough."

"You're not helping."

"Oh, but I am. I'm a doctor, you see. Distraction is part of the cure. I'll have that hook out in a jiff. But you'll need to lay off the çhilli vodka for a while."

"You don't say! As if I'm going to risk an 'afterburn' trip to the loo, though I will need a drink when this is over."

"You want my help, then?"

"Yes. Go on. Do your worst."

Courtney held my pecker between finger and thumb and then, in a fast and sharp twisting action, grabbed the hook and surrounding clothes and yanked it free from my manhood.

The pain was excruciating but short-lived. I looked down. My 'pride' was intact.

"There you go," he said, "I told you it would be easy. Now you can get back to your fishing. But if you do, you need to know the rules."

"What rules?" I replied.

"Firstly, that deliberately foul-hooking something is not sporting. And secondly, there's a size limit on this river. Eleven inches is a keeper. Sadly, the one you just had was *way below* that. It will have to go back…"

It seems that, when it comes to competition, size is important, after all.

JUNE

XII

A TROUT, FROM THE KIDDERMINSTER?

Each of us has our embarrassing stories. I've just shared mine with you. Thankfully, my 'pride' is healed and back to normal. So you can relax now. There will be no more tales about man-on-man tugging. Though I do tend to get 'caught up' in situations I sometimes regret. For example: do you remember The Flyfishers' Club that I mentioned a few chapters ago and its publication for which I write? Well, the Editor of the Flyfishers' Journal – Theo Pike – knows how to get the best from his writers. He encourages me to explore new and daring things, moving beyond my 'local' gaze to investigate the untried, unusual and unknown. Most recently, he's got me writing about the crossovers between bushcraft and fly-fishing, to build a series of articles themed around 'back to basics' fly-fishing. In it, I tie flies from whatever I find at the waterside and make tackle from natural materials – such as rods cut from a hedgerow, lines braided from nettle fibres and hooks made from thorn and bone. I even make flies from birch tar, peeling birch bark from fallen trees, charring it in a metal tub placed into a fire, and extracting the hot tar in a tin can. It's resulted in some innovative flies, but I scalded my fingers so badly making the tar that I had to delay writing an article by a week. It was the one and only time I've missed a deadline. Theo, it turned out, was not without revenge.

Known for his book *Trout in Dirty Places*, Theo knew exactly what punishment to inflict. "The dust will not settle," he said, "until you have caught a trout from *The Kidderminster!*"

'The Kidderminster' is a carpet-making town twenty miles southwest of Birmingham. It's built upon the banks of the West Midlands Stour – the one near to my childhood home that was so poisoned by carpet dye that its colour would change from red to orange to green to purple to blue. Theo knew that I had grown up near to the river, that I was one of the young urchins who would throw slices of white bread into its frothy water to see how quickly they would change colour, and that I would never, ever, have considered fishing there. The river, like the town, was a stinking, oozing, industrialised, nature-starved and unpleasant place to be.

Kidderminster, at least the town I knew in the 1980s and '90s, deserved its reputation for being "A nice place to drive past". Indeed, characters in the anarchic BBC sitcom *Bottom* named their toilet 'The Kidderminster' because, as they correctly stated, "It's the biggest ****hole in the world." What prospect, then, for a tweed wearing, cane wiggling, wild places loving angling purist to fish there? How could I write a bushcraft fly-tying article at a place where the only bushes would most likely be lying on the floor of the town's waxing parlour?

"Emphasise the urban environment," said Theo. "I'm sure you'll find plenty of manmade items from which to craft your flies."

I took a deep breath, put a clothes peg on my nose, grabbed a large bottle of industrial-strength hand sanitizer and headed back to the town of my youth.

Surprisingly, when I reached the town's ring road, there were no burnt-out cars, no chain-link fencing or razor wire barriers, no signs saying 'Keep Out – Danger of Death!', no knuckle-dragging tattooed Neanderthals, no smashed telephone boxes or torched speed cameras; just a normal-looking industrial town with rows of terraced houses and the occasional large chimney punctuating the skyline. True, many of the shops were boarded up, the town's leisure centre had closed, shell suits and baseball caps were still in fashion, multi-seat pushchairs were parked outside the town's college and riot vans were patrolling the high street, but Kidderminster was otherwise vastly improved since my last visit. Things were looking up for the town once contaminated by the greed of industry.

Theo had told me that the local council had regenerated an area that once housed the carpet mills.

This had become, he assured me, "a nice place bustling with shops, supermarkets and department stores. Flowing through the middle is the River Stour, newly gravelled, contoured and replanted." Further research revealed that, following years of improvement work by the Environment Agency, the old 'stewer' had been highlighted as one of the ten most improved rivers in the UK. A salmon was caught there in 2003 and, after stopping the release of permethrin (a moth-proofing pesticide) and Lindane (a carcinogenic insecticide) into the river by the remaining carpet factories, invertebrate life had returned. The food chain now sustained barbel, chub, dace, gudgeon, perch, and – ray of sunshine from the gods of angling – trout. A two-pound brownie had been caught from a ten-yard section of millstream that flows between the Town Hall and Library culverts. How's that for a return to form?

I drove to the rejuvenated area, parked in the supermarket car park and walked to the restored river. It was a wholly inviting and rather wonderful stretch of water. Fast-flowing runs tumbled across gravel and through lush ranunculus streamer weed and into deeper pools fringed by willowherb and willows. If I closed my ears to the sound of buses, cars, and 50cc mopeds, and averted my gaze from the rows of shops and warehouses, I could have convinced myself that I was standing next to a trout stream anywhere. It was intimate, clean, lush and fertile. Exactly the sort of river that screams 'trout!'

to those with a trained eye and hopeful heart.

Gazing down from a footbridge that crosses the river, I peered into the water and saw several dace of about half-a-pound and four chub of around twelve ounces. They were feeding mid-river on small nymphs drifting downstream. 'The Kidderminster' was no longer the butt of crude jokes and tourist map misdirection. It was renewed.

"Maybe," I thought, "I should cast to these fish?" No. I was to angle for trout. My Editor had told me so. And besides, I didn't have any flies with which to fish. I'd have to take a closer look at the riverbank to find some fly-tying materials.

Surprisingly, the materials I expected to find – litter – were pretty hard to come by. It would have been easier to make nymphs using willow bark and cordage from nettles and willowherb than from man-made items. But I had a specific challenge to complete. So I kicked around in the undergrowth, hoping to find things that would float, sink, glint, or bind.

Cigarette butts were plentiful. They could be used for the bodies of sedges or buoyant 'posts' of shuttlecock emergers. They could also be fluffed up and used as dubbin for a GRHE-style nymph or twisted into a wool-like thread to make a Baby Doll lure. I also found crisp packets and chocolate bar wrappers, all having brightly printed outers and shiny foil inners. They could be cut thinly to make ribbing, or twisted or plaited into shiny and lightweight wire substitutes. A polystyrene box, presumably from a local kebab shop, had multiple uses, especially for making floating beetles when coated with birch tar; and a bin liner ripped from a riverside dustbin could be used in the same way as the crisp packet. Best find of all, given that the fish were feeding on submerged items, was a handful of dried chewing gum that I was able to scrape from the pavement between car park and river. This could be moistened in the river and warmed in my hands to make pliant enough to be moulded onto the hook to make maggot-like grubs that would require no whipping. Good thing, too, as I was unable to find any form of thread. I'd half-expected to find some discarded fishing line that could be used to whip my flies. But there was none. I'd have to resort to once again using the tippet from my leader or a loose fibre from my sock.

Returning to the car, I made up half-a-dozen flies from what I'd found. The patterns would enable me to fish on or below the surface. I favoured a dark nymph, given my earlier observation, so I tied a bin-bag nymph to my line and went in search of a fish.

I donned my waders then climbed down the riverbank next to the car park. I stepped into the water. My waders did not fizz or start to dissolve, nor did they change colour. The water, which was cool and odourless, just swirled around them playfully.

A TROUT, FROM THE KIDDERMINSTER?

I began working my way upstream, flicking short casts into fast water and then longer casts into slower water below a road bridge. With Kidderminster College towering above and to my right, and a gang of youths threatening me from my left, I inched forward with a look of determination and sense of irony. The Black Country lad was back, fishing in a beautifully restored river, using a fly made from a bin bag. How my earlier self would have laughed.

Emerging from beneath the road bridge, I spied the entrance to the culvert that once held the two-pound trout. Coffee shops and bistros had been built above it, and the river frontage made this area popular with shoppers and diners.

"I'm fishing in front of a grandstand of onlookers," I thought, "so I'd better not botch my casts!"

As I waded around a bend, I glimpsed a rising fish. It was way upstream, alongside a willow copse, but feeding steadily and gently. The dimples caused by the rises were small, possibly from a dace or trout, but enough to make this fish worthy of pursuit.

Not wishing to spook the fish, I made long casts until my fly landed on the taking spot.

Nothing took on the first attempt. Or the second. Or the third. I kept casting to the spot, but without success. Rather annoyingly, the fish continued to rise next to my fly. I switched to a size 14 Fag-butt Emerger. It floated beautifully past a splash to its left and then, on the second cast, a splash to its right. Determined to catch this fish, and with a dozen spectators ushering me forward, I waded closer to the fish until I was level with the willows.

Then a moment of 'awk-ward' horror. There, partially hidden by the willows, was the source of the 'rises': a young couple engaged in an amorous embrace. She, a rather obese and greasy-haired teenager, was kneeling on the riverbank while eating a bag of chips. Her scrawny-looking suitor was hunched behind her, clinging on to her 'substantial circumference' while doing his best to eat her neck and left ear. The girl, totally unimpressed by her admirer's advances, was rummaging through the bag of chips for more satisfying fulfilment. Rejecting every third or fourth chip, she threw them into the river – making the dimples I'd been casting to – before chomping through whatever was left.

I, standing in the water looking like a voyeur with a riverside chip fetish, wondered whether the crowd had known this all along? Were they taking bets as to how close I would get or how long I would ogle the 'deep fried' coupling? I had two options: retreat discretely, enduring inevitable mocking from the crowd; or stay put until such time that I could offer the couple a handful of post-coital chewing gum. Of course, being a gentleman (and not wishing to sacrifice my hard-found fly-tying materials), I reeled-in and made my way back downstream.

A TROUT, FROM THE KIDDERMINSTER?

I'd failed to catch a trout but succeeded in rediscovering the 'youthful pleasures' of a restored river. It was a day of relationship building, for me and the young couple, and – as I climbed free of the water – I knew one thing: the dust had not settled. I'd have to return. At least, once I'd disposed of my litter, washed my hands, and eaten a very large bag of chips.

Stop – Unplug – Escape – Enjoy

How might you tie
and fish with urban flies?

June

XIII

TROUT FROM THE HILLS

> *"All the loveliest streams come down out of the hills, pouring over boulders, tumbling and churning among rocks and stones. All the bright and handsome little trout that inhabit such waters are there to lure a youngster away from roads that lead to towns and other ugly places."*
>
> Ian Niall

Fishing: it's a form of hunting which, as with all forms of 'seeking food for the table', is born of hunger. At least it used to be, before Man learned to farm food or a spotty teenager named Eric realised that he could earn a living delivering pizza by moped. My bushcraft fly-fishing articles were a way of retracing silken casts to a time when fishing required a greater deal of resourcefulness and ingenuity than it does today. (Even if that quest led me to a chip-munching, blubber-loving courtship ritual and the need to escape the jeers of a grandstand of onlookers.) Really, though, the articles conveyed my desire for escapism where

I seek a more authentic, organic, and slow-paced existence.

I have a theory: that people fish because they are searching for something. Often it is not for a fish. The question is whether the 'something' is that of our past, or future? Are we seeking to recreate something that we experienced before, or something new that we've yet to discover? I'm a traditional pleasure angler, one who uses vintage-styled tackle, sticks to traditional fishing seasons and prioritises relaxation by the waterside over the all-too energetic act of pursuing fish. Often, I'm accused of trying to recreate a rose-tinted view of the past that never existed. This is partially true. My form of angling is a nostalgic recreation of a bygone era, with rules made to suit those with a slow and loving heartbeat. But it isn't authentically vintage; I'll happily use modern hooks and lines and return the fish I catch. Traditional angling is a retro sport driven by image and artful enquiry that seeks to 'feel' the experience of being amongst nature. In doing so, it escapes the competitiveness of modern angling. Conveniently for me, it also suits my constant desire to slow down and taking things easy.

When it comes to trout fishing, there is less distinction between the traditional and modern angler. It's a branch of angling where, I'm pleased to say, the sporting ethic is strong. Fly-fishers know that they could catch more trout on a float-fished maggot, but they choose not to. They wish to be more sporting to the fish and outwit them on the fishes' terms. What, though, does this have to do with the quote at the start of this chapter?

Firstly, Ian Niall is my angling hero. His book *Trout from the Hills* is a classic. Whilst many fly-fishers drool over the writings of Halford and Skues, I prefer the down-to-earth and adventurous writing of Niall. He was a man who understood the meaning of the term 'wild trout' and fished as much for the sense of freedom as he did for the trout he caught. His adventures in the mountains of Snowdonia perfectly describe just where fly-fishing can take us: higher and higher until we reach our angling heaven.

Secondly, inspired by Niall, I'm about to embark upon a fishing adventure to the Welsh mountains. It will include all the essential elements: mountain range, llyns, a tent, the trout, vast skies of freedom, isolation, good weather, no midges (hopefully), and me. All being well I won't see another person for two days. I will fish before breakfast, hike through the day, and fish again at dusk. I shall sit out at night and witness the fullness of the night sky reflected in the vastness of a mountain lake.

Thirdly, I know that this style of angling is most personal to me; I enjoy it more than any other, as it's where and how I learned to fish. I've come to realise this after a twenty-year detour in search of other fish. Or, specifically, another type of fish.

American fly-fishing author John Gierach wrote,

"If you wanted a fish that could sip white wine and discuss Italian poetry, you'd look for a trout. If you needed a ditch dug, you'd hire a carp." Well, I've spent the past twenty-years searching for ancient strains of angling's ditch-digging brutes, this year discovering a lake atop a mountain that contains wild carp that have been undisturbed – I believe – since medieval times. I've written a book about it, so won't labour the journey here, but now that the quest is complete, I find myself yearning for a fish that's read Dante's *The Divine Comedy* and can sip a chilled Sancerre.

I'm craving an experience, one that involves trout from the hills, an adventure far from home, an opportunity to relax after a long quest, to do something strenuous yet relaxing and seek out the 'newness of the familiar'; to 'boldly go' where this man has gone many times before. It will be like blending the energy and wonderment of youth with the wisdom and gentler pace of adulthood. For I seek a very specific type of fly-fishing experience; one that balances the pressures of a grown-up life and releases the inner child.

Do you remember when everything was new, when a little baby could have 'the whole world in his hands' and the horizon was much closer? When everything out of reach was an irresistible frontier of adventure? When free time was infinite and 'responsibility' was dismissed as an unfounded rumour? When you could spend all day dreaming, playing, or observing;

when hours spent fishing would shape the rest of your life? These childhood times for me were spent in and around the lakes and reservoirs of the Plynlimmon mountain range of mid-Wales. They were relatively close to my West Midlands home and yet far enough away to remain 'mysteriously familiar'. It's where I had my every family holiday until I was twenty years of age. It was 'over the hills, but never far away'.

Too many people, in my opinion, travel too far in search of something they think they can't get at home. Like rocket-powered tortoises they jet off to 'concrete and sand' resorts where they cook on Gas Mark 6 for a fortnight before returning home looking like overdone Christmas turkeys. Sadly, they know little of what exists right under their noses at home. But while they're away, domestic holidaymakers like me enjoy trips to places like Wales where we sit in a caravan, watch the rain, read books, play board games, tie flies, and then, when it dries up outside, go fishing in the mountain lakes and get eaten alive by swarms of killer midges. It's what I call a 'proper holiday'. Real, life-affirming stuff that makes aficionados like me go back year after year to learn more about the things that seemingly never change.

Ian Niall says, "Some climb mountains simply to say they have climbed them and a few, a very few, look for yesterday and a sort of timelessness that is to be found in places that haven't changed since man walked the earth." The trout from the hills, the rugged landscape and the unpredictable weather provide the much-needed continuity in an angler's life. It's an opportunity to leave all the clutter of behind and rediscover the simple, pure pleasures of fly fishing for wild trout in wild surroundings.

It's time to pack our rucksacks. The experience is calling. We're going on an adventure.

JULY

XIV

REWILDING FLY-FISHING

Back to Basics

Let's do something radical and reimagine fly-fishing's potential to enhance the quality of our lives. In doing so, our innovative perspective should centre on Three Fundamental Facts:

Fact 1: Fishing enables us to get closer to nature. Different styles of fishing encourage this in different ways. Sedentary coarse fishers, sitting quietly while waiting for the fish to discover their baited areas, encourage nature to come to them. In contrast, the fly-fisher enters the fishes' world by matching the hatch, travelling light and venturing further into wild places.

Fact 2: Fishing unlocks the primeval hunting gene. Catching fish reassures us that, should we be called upon to put food on the table, we can feed ourselves and our family. In doing so, we maintain our status as the Alpha Male and can beat our chest in defiance of the Bargain Bucket and Drive-Thru culture of convenience living.

Fact 3: Fishing encourages escapism. The deeper we travel into the natural world, and the greater the number of technological encumbrances we leave behind, the more likely we are to escape the fast-paced lifestyle and stresses of the 21st century.

Sadly, modern fishing is not the simple and unburdened tiptoe into nature that once it was. The sport today is fragmented: split apart by conflicting dogmas, specialisms and unnecessary competition. It's mostly artificial, too, with coarse fish that are conditioned to eat human food (luncheon meat sandwich anyone?) and stocked trout that are as wild as a Chihuahua sitting in a pink velvet handbag. But nowadays, the opponent isn't so much the fish as our fellow anglers. It isn't enough to catch 'a' fish; we have to catch more and bigger fish than our neighbour. The pressure's on to compete and, in doing so, feel the need to buy more tackle and sundries that fund the industry that tells us to be more productive in our fishing. There isn't time to relax amongst the flowers. All that wildlife stuff, we are told, is just a distraction from our primary purpose: to catch fish. If we fail to land an ever-increasing bag of fish, then we fail as anglers. To succeed, we must use the latest tackle developments, tactics and fly patterns, just as every generation of fishermen has done before us. But in doing so, we get further from the Three Fundamental Facts of angling.

In today's mad, fast-paced, world, the pressure put upon us to be successful makes our recreations appear more business-like than perhaps they once were.

(Do you have goals and objectives for your fishing? Have you translated them into strategies and tactics for each water? Do you have an end-of-year review with the river keeper as you submit your catch return? And if successful, will you allow yourself a bonus for good performance?) Heaven forbid the day when fishing becomes the cause of our stress and we go to work for some respite.

What would happen if we stripped back our sport to its most primitive form? Perhaps by selecting the simplest style of angling – fly-fishing – and returning it to its origins? What would it reveal to our fellow fly-fishers and – more extremely – to the wider fishing community if we left behind all gizmos and creature comforts and lived feral for a week beside a remote lake somewhere? With just a rod, reel, line, leader, hooks and fly-tying thread, plus some survival items and a great deal of inner spirit, would we be able to rediscover the pure and original pleasure of fishing? Would the Three Fundamental Facts take on the brilliance of truth as we sought to catch fish to stay alive? And if we made it through the week, would we be able to shake hands with our primeval self? Sounds like an experiment worthy of our attention. But I wouldn't recommend doing it alone. Safety must come first until we learn the skills needed to survive.

The experiment could work if we got the right team together. We'd need people experienced at fly-fishing, bushcraft and campcraft; and perhaps someone who's good with a camera, and another who's good with a pen who could document the whole adventure. Fortunately for us and this madcap idea, I know just the people.

Thom Hunt is one of the UK's leading authorities on wild food. He owns 7th Rise, a rewilding centre in Cornwall and is skilled in immersing people in a wild environment. Mark Aspell and Manse Ahmad are owners of the Wilderness Pioneers Bushcraft School near Oxford. They are experts in primitive skills such as fire-starting, camp building, making tools and equipment from natural materials, and – crucially – know how to purify water using natural means. Nicky Brown runs Wilderness TV – a production company specialising in fishing programmes. He has travelled all over the world in search of adventure. And I'm an author known for his desire to get away from it all. We're all fly-fishers who seek to catch wild trout in wild places.

A number of phone calls were made. Each of my A Team accepted the challenge of a 'rewilding flyfishing' expedition. Thom agreed to lend his foraging, hunting and cooking skills to provide our meals. He'll share his knowledge of 'rewilding' people in wild places to spark plenty of intellectual debate around the campfire. Mark and Manse will be in charge of base camp and provide us with heat to cook and stay warm, shelter to stay dry, and clean water to wash our electric toothbrushes. Nicky will document everything with his camera, and

I'll be the one keeping a written log of events. We'll fish in rotation, with two people hunting and foraging while the other two rest and keep an eye on the campfire (maintaining a steady supply of boiled drinking water can be a full-time activity). And to uphold the survival element of our adventure, we'll tie our flies using whatever materials we can find. With luck, we'll be able to experience a pure version of fly-fishing. All we'd need to do was find the right location, pray for good weather, and tempt some obliging trout.

Lone Vigil

My plan to bring together a team of survival experts did not work out. When the time came to load up our cars and head into the wilderness, other commitments got in the way. All of the team were tied up filming elsewhere for TV. This left me feeling like a television widow, shameful of my boast that I could survive alone in the wilderness. But, the promise of adventure being what it is, I decided that I would head into the wilds. Alone.

My chosen location was the mountain range surrounding the Teifi Pools in mid-Wales. This is wild country, a landscape of mountains, moors and llyns covering 400 square miles. It's barren and bleak, but with a wild charm that screams "Freedom!" The fishing is accessible by car but, as the ancient Greek philosopher

Aristophanes wrote, "Man cannot discover new oceans unless he has the courage to lose sight of the shore." So, after three hours of driving from Oxfordshire to Ceredigion, I ditched my car in a farmyard next to the main road and hiked seven miles to the lakes, ensuring the time spent walking freed my mind from the distractions and disinterests of a modern life.

My equipment for the trip consisted of an eight-piece travel rod, a reel loaded with 6-weight line, three spools of tippet, two packets of hooks (size 12 and 14), a knife, a folding saw, a fire steel and tinder box, a waterproof poncho (that doubles as a tarpaulin), thirty feet of parachute cord, a rucksack containing a saucepan and Kelly Kettle, a bedroll and a sleeping bag. Oh, and a camera, writing pad and pen. It weighed little and I felt a sense of Boy Scout invincibility as I hiked up into the mountains. But I was hopelessly underequipped and – contra to the Boy Scout motto – was desperately unprepared.

This, after all, was Wales: where three metres of rain can fall annually. While walking towards the lakes, with rod in hand, I felt like a child carrying a kite towards a tornado. Water oozed up from the peat with my every step and gale-force winds pounded me from all directions. By the time I reached the pools, my poncho was doing little to protect me from rain that was 'falling' horizontally and stinging my face.

I managed to find shelter in the lee of a cliff adjacent

to Llyn Hir: a long thin lake in the middle of the cluster of pools. Here I had two priorities: get warm, then get fishing. The first would require a makeshift camp and fire; the second would require me to search for some fly-tying materials. Given that it was raining and that every ounce of firewood would be sodden, I opted to search for 'feathers and fur' which, by keeping me active, would also keep me warm.

An hour of searching produced three usable 'finds': a handful of sheep's wool found hanging on some barbed wire; a crow's feather that was floating at the edge of the lake; and a foil crisp packet discarded by a hillwalker. I had intentions of rolling the wool into a thread and creating a 'killer bug' style of lure, or combing it straight with a stick and creating a white-haired streamer; or cutting the crisp packet into ribbons and using it as tinsel or silver ribbing; or using the crow feather as a hackle to make a crude 'Black Spider'. These sounded great in theory, but with cold wet hands and no fly vice? Options were limited. Doubly so when I realised that I'd forgotten to bring any fly-tying thread. But with ingenuity and a minor sacrifice, I soon had plenty after cutting and unravelling the thread from my right sock.

Pushing a size 12 longshank into the cork handle of my rod, and holding the rod between my knees, I was able to roughly whip some strands of wool and finely-sliced slithers of crisp packet onto the hook. It wasn't pretty. It wasn't matching any sort of hatch. But this 'albino Christmas tree' was my way of making the best of what was available. Soon I had three usable flies and a very floppy sock. I went straight to the downwind end of the lake and began casting.

Nothing. I repeat, *nothing* happened. I fished for six hours, casting and walking; casting and cursing; walking and cursing; cursing and cursing. Not a pull, not a follow, not a fish. But then, on a blind and hopeless cast to the middle of the lake, the fly was snatched before I began to retrieve. The fish leapt from the water, pulled hard to my left, and then was gone. I was left nursing a limp line and an impotent ego. I fished on for another two hours, but without success.

A New Way of Seeing

Battling the elements was exhausting to the point of bone-numbing weariness. The biteless hours, with my 'head down' focus on catching a fish, made my heart sink. Ultimately, I found myself not wanting to fish. Rather I just wanted to be dry and warm. I retreated back to the lea of the cliff and made camp: a basic lean-to tarpaulin that flapped around flimsily in the savage wind. I crawled under the tarp and then into a soaking wet sleeping bag. Cowering under the flapping nylon sheet, with no fire or dry clothes to warm me, and only the haunting whispers of yearned-for food for company, I lay there shivering and feebly trying to breathe warmth into my hands. Anywhere seemed more appealing than this barren, soaking, windswept winterscape of a summer mountain. I was reminded of the words of Thoreau, who said: "We should come home from adventures, and perils, and discoveries every day with new experience and character." Wise words. Mine was an experience and although the endurance was character building, I found myself wanting only to 'come home'.

I'm an adventurer, at least in spirit and optimism. But this situation was real and way more than the experiment I'd bargained for. It was teaching me a valuable lesson: that while I'd sought to understand the relationship between fly-fishing and the wild, I'd quickly learned that fly-fishing consumes too many calories to make it a viable survival technique. Primitive man set handlines at the water's edge while foraging or hunting elsewhere. He was smart, whereas I, the evolved man of lesser knowledge, sought some sort of pseudo recreation. I was 'playing at it' in an attempt to connect with the wild. (I wonder what my primeval ancestor would have made of me, feeling thoroughly sorry for myself, after less than a day exposed to the elements?)

Yet, with some degree of bitter irony, it was not Nature but Man who had enabled me to fish. Neither the sheep, barbed wire, crisp packet, or nylon thread from the sock were native to the mountain. They were put there by humans. Artificiality had enabled me to connect to the one thing that was truly wild: the trout. So perhaps Man's impact on the wild isn't that bad, after all? (Doubtful, given the extra vegetation and shelter that would have been present in the absence of sheep.)

The thing that the experiment taught me, above everything else, it that being alert – through terror or excitement – and responding to the intensity and challenges of life that exist in wild places – is the essence of the wild. It's the far-from-delicate balance between what we fear and what we relish: the heartbeat that quickens when we're forced to live in the moment. Rewilding the human spirit can be achieved through fly-fishing. It's the view from a mountaintop, the sound of woodland at dawn or a rise to a fly, that calls to us with ferocious sublimity, quickening us to be real.

We dig deep to discover our treasures; seeing their true beauty only when we lift them to the light.

July

XV

TROUT AT TWO THOUSAND FEET

"A mountainous wilderness extended on every side, a waste of russet-coloured hills, with here and there a black, craggy summit. No signs of life or cultivation were to be discovered, and the eye might search in vain for a grove or even a single tree. The scene would have been cheerless in the extreme had not a bright sun lighted up the landscape."

George Borrow

"There's gold in them hills," said Uncle Bill as he drove me in his car through the mountains of mid-Wales, "but you gotta dig deep." My uncle, an ex-miner, knew of the dangers of tunnel mining. His respect for the mountains, and those who explored deep within them, could be heard in the trembling of his voice. I gazed through the car window at the scree-scarred mountains and imagined the generations of Welsh miners who had 'dug deep' into the mountains and into their courage to brave the depths of the underworld. It was enough to inspire me, as an innocent ten-year-old, to stare at the mountains and imagine what lay within. Twenty-five years later, I'm still imagining what lies – and 'breathes' – within the Welsh mountains. They always make me stop, stare and lose track of time. Of course, these days I know that the mines in that particular area of Wales (Cwmystwyth and Rheidol) produced mostly silver and lead, but it's never stopped me dreaming and knowing that this area of Wales is golden.

Up in The Mountains

I'm writing this while sitting on my coat and leaning against a dry-stone wall next to Nant-y-moch reservoir in mid-Wales. I'm about twelve miles north of the Teifi Pools, where I 'rewilded' my view of fly-fishing, and I'm enjoying an altogether warmer and more hospitable experience. I've found a sunny, sheltered spot in which to reflect and write. I am, most assuredly, in my chosen element.

I'm gazing out, across waters that are swept and patterned by a warm south-westerly breeze, to the Cambrian Mountains beyond. They 'loom high and reflect low'. Drosgol is opposite at 1,806 feet and Banc Llechwedd Mawr is in the distance at 1,837 feet. They look ominously close, yet Llechwedd is several miles away. I know this because my Ordnance Survey map says so. I also know that the grandaddy of them all – Pen Pumlumon Fawr (also known as Plynlimmon) – is just out of sight to my right. At 2,467 feet it is the highest of the Cambrian mountains; it gives birth to the rivers Severn, Wye and Rheidol, is home to red and black grouse and some of the best wild brown trout fishing in Wales.

I came here following a hasty 'ponchoed retreat' from Llyn Hir, seeking safe territory and easier fishing – this being one of the reservoirs I'd fished since childhood. I'm here on different terms, having to decided to abandon the survival camping trip. Whilst a soaking wet sleeping bag and pan-fried earthworms might once have appealed, my recent survival ordeal made me realise that the reckless spirit of youth had departed my bones. My sense of invincibility had been replaced by a spindly-legged and saggy-bottomed wisdom that seeks out easier ways of doing things. So, instead of continuing the wild camping, I booked myself into a farmhouse B&B and relaxed at the prospect of a warm bed and a 'full Welsh' breakfast. But as the day was young and the weather had improved, I'd also decided to enjoy some traditional, classic, fly-fishing. Just a few hours. Before the pub opened.

Nant-y-moch, in case you think I'm wimping out, is not an easy water to fish. It's huge: 680 acres of Welsh mystery that fills two valleys and conceals a flooded hamlet and numerous Iron Age settlements. It has a twelve-mile circumference and contains over

26,000 million litres of water. And, as it's not especially rich in food, it contains mostly small fish – and not very many of them. It's 'beautifully compelling' but tough-going. It's not for the faint-hearted. But as with so many vast reservoirs (I'm thinking of the likes of Grafham, Blagdon and Chew) it's best to think of them as a series of interconnected lakes, each with their own bays, promontories, inlet streams and wind lanes. Time spent watching the water soon reveals where the fish are rising, which is often within a few feet of the bank. A small fly pattern, with a touch of red and/or silver and employing a hen hackle, moved slowly to give a jerky movement, will usually give results.

"That's all very factual Fennel. Are you sure that you didn't steal that guidance from an information sheet for the fishery that you acquired in 1981?" – *Yes, but being a non-conformist, having absolute faith in one fly alone (the Hudson Sedgetastic, more of which later) and being undaunted by a challenge, I would have to ignore the fly patterns recommended by the locals, tighten the laces on my walking boots, and do things my way.*

Having checked in to the B&B and laid out the previous days clothes to dry, I jumped into the car and headed up into the mountains towards Nant-y-moch. The road to the lake was more potholed than I remembered, and the surrounding moorland was boggier and mistier from recent rain than I'd have liked, but it was reassuringly inviting – like a craggy, unexpectedly damp, and slightly vacant old relative.

I parked the car in a layby on the right-hand fork of the reservoir, near to where a stream flows in. I exited the car, collected my fishing tackle from the boot, then walked casually towards the water. I got fifty feet into the soft rush and heather marshes before breaking through a crust of peat and sinking knee-deep into peat-slop. I didn't mind. The water before me (and in my boots) had a good feel. I was part of it, in my soggy-socked and 'squelchily idealistic' way, and soon would seek to connect more completely by catching one of its trout.

I pulled myself up out of the bog and, walking like a heron in lead boots, moved forward while keeping one eye on my footing and one on the water before me.

Out there, just ten feet from the bank, were the circles and bow waves of feeding fish. Trout. Trout! Right there! Feeding!

I crouched down low amongst the rushes and sedges and tackled up my rod. I selected a size twelve Sedgetastic deer-hair fly on a ten-foot leader and then crept ever so slowly to the water's edge. I flicked out the line, barely having to extend any fly line from the rod tip, and saw the sedge land gently on the water just to the right of a rise. Within seconds it was engulfed in a bold swirl. I counted 'one-two' and then struck, feeling the jagging, juddering sensation of a small trout on the line. It darted about and leapt clear of the water before I drew it quickly to my hand. It was a beautiful pewter-and-bronze-scaled and red-spotted brown trout with fins the colour of wet slate. At nine inches long, I'd say it weighed about six ounces and was average for the reservoir.

I caught a further three fish from that area before retiring back to the car to get my Kelly Kettle. I filled it with water from the reservoir, loaded its fire base with heather that had dried in the breeze, then set light to the kindling and watched the smoke rise from the chimney. It drifted gracefully out onto the reservoir, partially concealing the images of fish and ripple, lingering just long enough to tease with glimpses of promised contentment. Sure enough, the resulting brew was the perfect end to an afternoon's fishing.

Higher than Heaven

As I sat beside the reservoir, drinking tea and reflecting on recent successes, my eyes became drawn to the peak of Plynlimmon to my right. I remembered the words of Hervey Voge, who said: "The mountains will always be there, the trick is to make sure you are too." So I finished my tea, packed my fishing tackle into the car, turned away from the lake and then set off on a hike up the mountain. I followed a track at first, which led me to Llyn Lygad Rheidol, the southernmost glaciated cwm in Britain. I walked to the left of the lake and followed a ridge towards the top of Plynlimmon.

An hour and a half later I was sitting on the peak of the mountain, breathless and overawed by the beauty of my surroundings. How far could I see? It seemed like a hundred miles, but there was Snowdon on the northern skyline, a mere forty miles away, staring at me in disbelief. "You're supposed to be fishing," it whispered. But no amount of angling could compare to the view before me. I knew there were trout below, but I stayed up there on the mountain for most of the afternoon. Eventually I descended back to Nant-y-moch where I sat and watched the sun sink behind the mountains. I stayed there until well after dark, marvelling at a billion stars overhead and their reflections in the water below. I stood quietly and motionlessly, thinking deeply about the majesty of the heavens. So much beauty obscured by daylight, revealed only when we 'step outside' and gaze in a different direction. It's what we can't see, as we gaze into the 'void between', that draws us in. Much like the act of fishing, when we cast into the unseen depths. Hoping, not knowing, that something is there.

Isolation

My time on the mountain, and beside the reservoir, reminded me of a quote from one of the first fishing books I ever read: Maurice Wiggins' *Fishing for Beginners*. Written in 1953, it reminds the reader of the emotional and spiritual benefits of the sport: "An angler's world is a world apart...It is a world of mystery, silence, and beauty, a world of unending adventure, and a world of solitude. Every angler's world is a private world...it is a great world you enter when you go fishing. A wonderful, beautiful, exciting world."

Fly fishing in remote places is liberating. There's just you, the water, the fish, and a vast sky. You don't have to look pretty in the mountains. There's rarely anyone around to see your false casts, trailing loops or line caught around your feet. You can catch the heather behind you and not feel a need to curse. It's as much about the place and the adventure as it is about the fishing. Being part of the landscape, seeing mountain hares lollop by and sheep walk past without bolting, being alone in the dark beneath a canvas of stars; it's life assuring and, I would argue, life-defining.

When I am too old to fish, or too weak to walk great distances, I shall think of my time in Wales and remember how the mountains were dwarfed by the starlit sky. The image will remind me of the old tourist advert for Alaska which said, "When you arrive here you feel really small; when you leave, you feel big." I'll also hear the words of mountaineer Rene Daumal, from his *Mont Analogue*, saying: "You cannot stay on the mountain forever. You have to come down again. So why bother in the first place? Just this:

What is above knows what is below, but what is below does not know what is above. One climbs, one sees. One descends, one sees no longer, but one has seen. There is an art of conducting oneself in the lower regions by the memory of what one saw higher up. When one can no longer see, one can at least still know."

It is with this 'knowing' that I ready myself to return to a world much smaller and busier than up here. But not before I acknowledge the man who guided me here for this adventure.

Stop – Unplug – Escape – Enjoy

How has fishing shaped your life?

July

XVI

ALONE BUT NOT LONELY

My time in the Welsh mountains, living wild (for a day) and fishing for more than just my supper, made me appreciate the potential intensity, majesty, and vulnerability of life. I'd connected with my wild side and that of fly fishing, and proved that angling can take us into the heart of nature. But, secretly, I was also trying to connect to something – or rather someone – I was missing.

The adventure had been made possible following a chance meeting with Moc Morgan (Wales' greatest fisherman) during the annual dinner of Tregaron Angling Association. I was there to give the after-dinner speech and Moc – as patron of the Association – was my host at the head table. He and I chatted about fly-fishing, writing and broadcasting, and how fishing helps to cleanse us of the stresses of life. We also discussed the current obsession with 'angling mechanics' (catching fish) in the angling press: "Just being in the countryside and being part of the pageantry of nature," said Moc, "is pleasure in itself."

I sat in awe of Moc's experience and wisdom, he

being forty-five years older than me, just listening to and reflecting upon his words. His final comment, for which he was known, was that when fishing we are "Alone but not lonely". Four words that summed up everything I seek when 'getting away from it all' with a fly rod.

As part of my speech to the Association, I'd commented that I've never had the good fortune to catch a sewin (sea trout). When I returned to the table and sat down next to Moc, he immediately offered to help me catch my first fish: "It is said that the sewin was the last fish to be created on that fifth day of creation," said Moc. "After ample practice on all other species, at last came forth perfection. You've had your practice. Now we will fish for perfection!" We agreed to meet upon the banks of the River Teifi in June, when we would sit together, drink tea and talk softly as we awaited the arrival of sewin into the pools. Alas, it was not to be. Moc died suddenly on the 25th May 2015, aged 86 – just a month before he and I were due to fish together.

Moc's passing was mourned by a nation of Welsh anglers and an equal number of friends and followers from elsewhere in the world. In my grief, I travelled to the Teifi Pools (controlled by Tregaron Angling Association and a favourite haunt of Moc) attempting to connect with the true spirit of fly-fishing and feel close to the master whose words had put fishing in Wales firmly in the hearts of fly-fishers everywhere.

ALONE BUT NOT LONELY

My calling came while watching Moc's 'Life in Fishing' programme, recorded for the Fieldsports Channel on YouTube. Filmed with his son Hywel Morgan, the four-minute piece was set at Teifi Pools and contained his Eternal words:

"I am in my earthly paradise now. This is where I come to escape from the madding crowds... You switch off when you're here. Alone but never lonely, because this is wild country and I'm part of it when I come here... Fishing is a breakaway to take you to a different world. And you must realise that this is religion with us – and not a sport. A way of life... The world turns around fishing, and fishing has given my life a purpose. But you really live for the time that you've got to spare – and the times you haven't got to spare.... God we've got a lot to be grateful for. And I thank the Lord that I have been able to spend my time and my life here. Alone but not lonely."

This is why I sought to be 'alone but not lonely', practising my religion in the same Welsh hills, amongst the same llyns, that Moc had called home.

You know how my story panned out, with me ending up as wet and cold as the fish. But this chapter is about Moc: the man whom I met only once but knew (though his writing) my entire life. It's about how I came to witness the unveiling of his memorial stone while standing proud, shoulder to shoulder, amongst his friends and family. Here's my attempt to celebrate the man to whom I am so indebted.

Morgan John 'Moc' Morgan was born on the 7th November 1928. He grew up in Doldre, Tregaron where he was mentored by renowned local fisherman Dai Lewis. Moc described him to me as "One of the giants of the fishing world, whose guidance when casting to a rising fish was 'never to cast above,' so the trout can get a good look at the fly; 'rather just to the side and close to its eye' so that it only gets a glimpse of the fly and instinctively rises." Moc explained that Dai had always insisted on casting accurately – "that you must be able to land the fly within two inches of your target", and Moc had championed this advice when coaching the Welsh Fly-fishing Team to their first-ever victory in an international competition. It was the 1967 International Championship and, Moc said with a glint in his eyes, "The one where we gave the English a right good thrashing on the way!"

A teacher by profession, Moc taught first at Pontrhydfenigaid School. He later took the headmaster post at Lampeter Primary School. Education was in his veins, even as an angler, and throughout his life young people flocked to him for advice. As his wife Julia explained to me: "They loved him, buzzing around him like bees to a honey pot. He always said that he learned as much from them as he taught them. He always spoke to them as equal, never raising his voice, always able to get them to agree by coming at things from a different angle. Confrontation was never necessary."

Moc's profile grew in the 1960s when he was invited onto a Welsh television programme to discuss his love of angling. This led to him hosting a show called *Country Life* on Radio Cymru, and later the television programmes *Moc's Country* and *Moc's World* for S4C. He wrote five English language fishing books, including *Fishing* (1977), *Fly Patterns for the Rivers and Lakes of Wales* (1984), *Successful Sea Trout Angling* (co-authored with Graeme Harris, 1989), *Fishing in Wales: A Guide to the Lakes and Rivers of Rural Wales* (1990), and *Trout and Salmon Flies of Wales* (1996). He was a longstanding columnist in the *Western Mail*, and wrote the regular Welsh river reports in *Trout & Salmon*. (He said to me when we met, "You *have* to keep writing. Why? Because the best fishing is often found in books.") His works led to him becoming the authority on fishing in Wales. In 1986, the former American president Jimmy Carter – a keen fly-fisher – employed Moc as a fishing guide when visiting mid-Wales. In 1991 Moc was awarded the OBE for organising the World Fly Fishing Championship, also in Wales. He was three-times president of the International Fly Fishing Association, secretary (and later chairman and president) of the Welsh Salmon and Trout Angling Association, managed the Welsh International Fly Fishing Team for nearly 30 years, was founding chairman and chief executive of the Federation of Welsh Anglers, and was head of Fly Fishing for the Disabled. He was passionate about getting everyone into fishing, so he set up the Welsh Youth and Women's fishing teams. His life was dedicated to fishing. And yet throughout it he remained grounded, irrespective of his achievements.

For over fifty years, Moc always began the first day of the trout season at Bont Llanio on the River Teifi. It was rather fitting, therefore, that his memorial service and unveiling of a stone in his honour was held at this location on Thursday 3rd March 2016: the first day of the fishing season. It was here that Moc had greeted members of Tregaron Angling Association, always saying a few words to commemorate the new season, while 'blessing' the river with a few drops of Penderyn Welsh whisky. The tradition was continued during the service by his son Hywel, who provided the blessing. Commemorative words – this time in Moc's honour – were said by Donald Patterson, chairman of Tregaron AA. Scripture readings were given by Rev. Phillip Davies; a tribute was paid by Rhys Llywelyn (Fishing Development Manager for Wales) and a vote of thanks was given by Eifion Davies (Natural Resources Wales). Finally, the commemorative stone and plaque were unveiled by Moc's widow Julia. The entire event was filmed by S4C.

With the exception of Donald Patterson's address, the whole service was conducted in Welsh. So, Englishman that I am, I cannot tell you what was said. But this didn't give me any less of an understanding of the love being shared for such a great and treasured man. All the inflections, tears and smiles when the speakers gazed skywards. I understood.

Thankfully, the order of service provided a translation of the inscription carved upon the plaque. It said:

"The silver river of Bont Llanio
Gilded by Sun Fly
Was Moc Morgan's heaven
Skilful Kingfisher."

A fitting tribute, which also acknowledged Dai Lewis' Sun Fly that Moc fished with all around the world, describing it as his "masterpiece of deception".

While speaking to attendees after the service, I asked them if they could sum up Moc in a single sentence. None of them were able to do so. But all of them paused before speaking of the great man. It was the pause that said it all. A reflection. A love. A loss. An emptiness. Julia's response was best: "I cannot sum him up, or adequately say how I feel, in English. The Welsh word is 'hiraeth' but it doesn't translate. Its closest English words are 'homesickness' and 'yearning', with a strong sense of rootedness, place and belonging. But it's much, much more than this – a real thing that crushes with its weight; it impacts one's existence, keeping us awake at night and controlling us by day. Moc is gone. Hiraeth is here."

There's no greater reason than life and death to speed

our heartbeat and our actions. We do not live forever. Places change. We may not always have the opportunity or strength to do the things we love. There is only the moment. We should grasp it and live in that moment.

As anglers we know that even if we find ourselves alone, we're never lonely.

September

XVII

THE NET ON OUR BACKS

Have you ever stopped to study your shadow while fishing? That presence upon the earth that leaves no trace and appears boldest when we stand tall against the sun? It's what I'm doing now. The shadow shows the rod in my hand, a fly reel tilted upwards and a net hanging from my back. What does it tell me? That I am an angler. Equally importantly, that I am doing what I most enjoy. But does it change when we stop fishing? Does the shadow move with the slowness of that on a sundial, or does it morph into another shape entirely? It's got me thinking about how and if our identity as an angler influences our activities away from the water.

Think about this example: a young man looks into a mirror and sees his older self. "Is that me?" he asks. "My, how I've changed. I barely recognised myself." Yet he is still the same man, with the same thoughts, values and emotions. His eyes are the same, but his face and body have changed. Now think of the mirror as a metaphor for the transition between fishing time and 'other' time: at home with one's family, at work, on holiday, at the shops, at church, down the pub.

Wherever. We might remove our waders and floppy hats when we cease fishing, but we're still the same person. How does that affect our behaviour in everyday life?

Ages ago, when I was a junior manager, I had an office with a lockable door. I'd hide myself away in there and read fishing books, tie flies, sleep (following all-night angling trips) and answer the door only when I absolutely had to. It was a luxury, being so 'unwanted' and easily forgotten that I could skive for days or weeks on end, but it was also a selfish and cowardly existence. I knew that once I opened the door, I would have to become someone different. I'd have to behave differently and, while acting the part expected of me by my employer, speak and think differently too. I didn't want this and I didn't want the world to see the real me. So I hid the real me away, behind the locked door that separates knowing and unknowing (and, most likely, prevented me from getting a disciplinary from my boss). But I did it as I was too shy to declare that I would rather be sitting quietly beside a stream, with fly rod in hand, than boasting of any amount of career achievements. The angler within me, therefore, was contained: hidden from view. My shadow became fainter and the net disappeared from my back. If I'd looked in the mirror, I wouldn't have recognised my face, image or behaviours. I was, when people saw me emerge from my office, 'industrially cold' and business-like; not at all like the meadow-skipping romantic that emerges when I'm fishing. I'd kept the real me – the one whose heart flutters at the first birdsong of dawn – safely protected behind a locked door. Why was this so?

It takes courage to stand up for one's beliefs and be proudly different, especially when expectations are that we should appear and behave differently. We might not be able to wear a fishing waistcoat to the office, but there's no reason why our values as anglers can't seep into every aspect of our lives. If we are an angler, then we are an angler. Period. We are not an employee, chief executive, car mechanic, or shop steward one minute and an angler the next. There's only one hat, and it's likely to have a Greenwell's Glory hooked into it.

Where, then, should we begin this transformation from 'adaptable' to 'authentic'? There's an old saying, usually attributed to Confucius, which says "Give a man a fish, and you'll feed him for a day. Teach a man to fish and he will eat for a lifetime." (Or, if your partner's reading this: "Teach a man to fish and you've fed him for a lifetime. Unless, of course, he doesn't like sushi – then you'll have to buy him a Kelly Kettle, hobo stove and frying pan as well.") I propose that the focus of this saying confirms the suggestion that anglers are a selfish breed, interested only in indulging their hobby and spending time on their own. Ignore the earlier versions of the saying and try this one instead: "Give a man a fish, and you'll feed him and his family for a day. Teach a man to fish and he can feed his family for a lifetime." Notice the difference? He's doing it for others as much as himself.

Wilhelm Stekel once said, "The mark of the immature man is that he wants to die nobly for a cause, while the mark of the mature man is that is wants to live humbly for one." Our noble cause is angling, but it teaches us skills that can be used elsewhere in our lives. For example: good observation helps us to spot feeding fish, it also enables us to identify opportunities at work and things for the family to do; perseverance enables us to keep fishing through harsh weather or lean periods when the fishing is slow, it also helps us to complete jobs at home and at work; patience leads to tolerance, which makes us more accommodating to others; a sense of humour helps us to laugh at our dry nets when fishing, it also enables us to cope when we've failed miserably at a task (and the wisdom to laugh when we've failed contentedly); the ability to plan for success helps us to deceive fish and learn from outing to outing, it's also is an essential life skill if we are to realise or dreams; and, most importantly, the ability to care – not just for the angling environment – should be something that's focused on everything we love. There are many more examples, when you add them up it makes for one conclusion: we should be proud to be anglers and use this to the benefit of the world around us.

Albert Einstein said, "A person starts to live when he can live outside himself." If you were part of the world looking in, you might not know about all the skills and values of an angler. Many see only 'lost souls' wasting away their lives in the freezing rain and returning home empty handed because they have returned everything they caught. They see hopeless people performing a pointless exercise and making little difference (sounds a lot like senior management). It's not true (for anglers at least). Anglers are custodians. We pick up litter when we see it, maintain waterways, report signs of pollution, fight the polluters, restore water habitat and encourage youngsters to take up a rod in comradeship. We represent the brotherhood of anglers who, together, stand taller than the individual and cast a solid shadow. We are the energy, the force, and the power to be reckoned with.

Hmm. Really?

Anglers fall into two camps: those who believe that if we want to get close to nature, then we have to stay low, out of sight, and down to earth; and those who know that it's questionable to assume that we are collectively strong when we as individuals are weak.

We have to live what we believe if we are to be true to the cause and support the brotherhood. I say, "Lift your fishing rod high and prepare yourself for the charge into battle. We've got to fight to protect and enhance our watery world." There are several groups of anglers already engaged in this fight. (The Angling Trust and Salmon & Trout Conservation being the main forces in the UK.) But I want to draw your attention to the wonderful work done by the Wild Trout Trust, in whom I strongly believe. I've supported them since their formation in the late nineties and have enjoyed seeing them roll up their sleeves and 'get on with it', improving trout habitats and enhancing the prospect and profile of trout in our rural and urban waterways.

The Wild Trout Trust's practical approach appeals to my active side. While there's an argument that the future of our trout rivers should be fought in the courts and in parliament, it's the hands-on people, waist-deep in water, who make the immediate difference. (Again, it comes back to the commitment and conviction of the individual and the collective purpose of the masses.) A four-pronged attack (covering politics, water quality, promotion and river management) is the best long-term solution for the future health of our rivers and sport, but think how effective your average working party is when more people attend and chip in. The strength of our collective back isn't the point, it's the courage to stand up for our beliefs and demonstrate what it really means to be an angler.

All this passion, just from catching a glimpse of our shadow. There's no such thing as 'me and my shadow', only the man who knows himself and understands his calling. As Audrey Hepburn once said, "It's that wonderful old-fashioned idea that others come first and you come second." So when you next gaze upon your shadow, don't be tempted to rise onto your toes to make you seem taller; just stay there and see your shadow grow as the sun sets. The longer you stand, the longer you are. You are an angler, which means everything.

Stop – Unplug – Escape – Enjoy

In what ways does being an angler define your identity?

AUGUST

XVIII

A GRILLING FOR A TROUT

An Indian Summer is here and with it a week of blue skies and sweltering heat. All around seems to be held in a state of 'poised transition', like a garden swing at its highest point refusing to come down. We know that autumn is on its way, but for the moment we can merely mop the sweat from our brow and tiptoe across the scorching earth, feeling somewhat like popcorn in a pan.

The countryman within me will say that this is a time of year to be out and about at dawn, when mist spills across the meadows and the air smells of dew and freshly cut hay; or to walk upon on the heath at mid-day and listen to the broom pods cracking open, or to find a shady spot beneath a riverside willow and dangle one's toes in the water. The trout fisher in me, however, knows that this is a time to be indoors with a cold beer. Trout, like me, are not comfortable in hot weather. They either sulk in the depths or leap about as if their brains have started to itch. They might feed at dusk, but during the heat of the day they are best forgotten.

I am not a sun lover. Being so fair-haired and pale-skinned means that ten minutes in these rays will make me redder than a flamingo that's been caught with its pants down. I am a man of cool days, when I can wear my felt waistcoat and tweed cap and step forth with English dignity. This might also explain why I've ended up standing in my garden wearing a knotted hankie on my head, trousers rolled up to my knees, and doing my bit for the male ego.

Mrs H has arranged a barbeque party for her girlfriends. Naturally, like most men, I volunteered, spatula in hand, to do the cooking. I am, after all, the Fire God; King of the burnt sausage and charred burger; man of many menus, from 'Cremation Chicken' to 'Cider and Cinder'. I'm the ultimate 'cordon noir' chef. Mrs H's Ladies don't know what's coming. They wave at me through the living room window as they sip Pimms and talk about making cupcakes. But I daren't go inside to warn them. (I made the mistake earlier of popping in for a beer just as they were talking about 'the squeeze-to-squirt ratio when piping a cream-filled bag'. Never again. I'd rather stay out here and cook.) Soon the ladies will be prodding their forks at lumps of blackened food and saying things like, "Yes, I think it's dead" and "You must give me your recipe some time." By my reckoning I've got about an hour of respect left before the mockery begins. Well, the barbeque is lit and there's enough smoke to conceal an escaping angler, so I could make myself scarce for a while. I could go fishing. I've just got to hope that none of the women notice I'm gone, else I'll get a proper grilling from she who chars the coals. (The balance of power in the modern relationship is worthy of further study, but as I'm short of time I'll simply say that I let Mrs H think she's in control by agreeing to and doing everything she says. Life's easier that way.)

The closest water to my house is the River Windrush, but the trout won't be doing much at the moment. My best bet is a lake about five minutes away in the car. I could go there, spend twenty minutes fishing, and then be back before Mrs H knows I've gone. Risky. Oh so risky. We'll need a plan. How does this sound: if I get caught out, or if Mrs H asks where I've gone, we'll tell her that I had to go and buy some more charcoal because the stuff we've got is for Australian barbeques and thus the flames were rising in the wrong direction? Yes, that should work. Girls don't understand such things. (Just like I had no idea what they meant by "Give it a poke down the centre to check it's risen properly" when I popped into the house earlier.) Men and women have a mutual ignorance that knows it's best not to ask about such things.

Okay, I'm off. I'll grab a rod, reel, waistcoat and net from the shed, and then race like a rally driver down the Cotswold lanes until I reach the lake. I have thirty minutes, start to finish, during which time I'll attempt to catch a trout. Right, I'm logging off now. Back in half an hour.

I'm back and keen to report on my adventure. The good news is that the trout were rising when I arrived. They were cruising about the middle of the lake at ultra-slow speed, sipping buzzers like basking sharks filtering plankton. I knew it would be easy fishing, but I had a challenge. In my haste to leave, I picked up the wrong rod. Instead of the intended eight-foot five-weight rod, I'd accidentally picked up my newly-restored nine-foot eight-weight sea trout rod that just wouldn't load properly with the five-weight line. As a result, each cast ended up as a coiled heap in the water. I just couldn't generate enough line speed to straighten the leader and properly present the fly. I ended up having to retrieve all the slack line and leave the fly drifting in the water just fifteen feet from the bank. I watched it drift slowly, all too slowly, towards a rising fish. After fifteen minutes I saw a trout approach the fly, rise lethargically, and porpoise back down into the depths. I waited three seconds and then struck. The fish woke up and ripped around like a caffeine-tripping student at exam time. Short of time, I hauled the fish towards me, netted it and dispatched it. The fish, a brown trout of about two pounds, was slipped into my creel.

That was eight minutes ago. I'm now back in front of the barbeque, with 'manly tongs' in one hand and pen in the other. I feel smug for escaping, and confident at being able to catch so quickly. It wasn't relaxed or romantic fishing. It was an adventure with singular purpose: to grasp an unexpected opportunity. And just in time, too, as Mrs H is calling me.

"What's that, dear?" I ask.

"I was calling you," she says. "I was wondering if we had a trout that we could put on the barbeque?"

"One step ahead of you sweetheart. I got this one from the freezer earlier."

"Oh, you're so considerate. Here: have a beer for being so lovely."

My, how I'll enjoy the ales consumed this afternoon.

Stop – Unplug – Escape – Enjoy

If you went fishing, right now,
how many fish could you catch
before you were missed?

October

XIX

CATCH RETURNS

The trout fishing season has ended. I should be sad, but I'm not. I'm stressed. Its demise signifies an awkward time for a traditional angler like me. Not that I'm worried about a season ending (seasonality in angling is what keeps our sport balanced and in tune with nature) but because it marks the time when I have to confess to how few trout I've caught this year and get statistical about my fishing. It is time to submit a catch return.

Thinking about how many, or few, trout I've caught in a season is no more meaningful than calculating how many times I've worn a favourite hat. The 'enter total here' box just isn't relevant or reflective of my season's enjoyment. Total what? I go fishing for reasons other than catching fish. In fact, I don't like trying to quantify emotion or rationalising pleasure at all. Doing so is like describing a sunset as '8.47pm'. That we smiled and have happy memories is enough. But obligations are obligations. The catch return must be submitted. I'm sorry, but I'm going to have to get serious. It's time to do some numeric reasoning.

Submitting a catch return is a complicated business.

One has to remember the exact number, species and size of fish caught; when and on what fly they succumbed to your rod; from which beat of the river they came and at what time of day; and in what weathers, atmospheric pressures and wind speeds you decided to take a pee. In fact, all the unimportant and boring things that can't determine whether you had a good season. But record them we must – to appease someone who likes statistics and most probably carries a tape measure with him when fishing. (I guess it's to 'measure his pleasure', done when no one else is looking. You can imagine it, can't you, "A-rise! Aha! There you are! I'm going lay you out, run this tape along you and then plot you on a graph!")

It's fair to say that presenting me with numbers is like asking Gandhi to bungee jump off an elephant. I can do it, but it's not my usual style. That said, I can add, subtract, divide, multiply and count to a thousand and thirty-seven. (Anyone who's counted beyond that must have been in serious need of entertainment.) Calculus and algebra were never my strengths at school and mental arithmetic was exactly that: mental. They might as well have been the names of psychotic French footballers. Which is why I need your help. I've only caught three trout this season, so I'm going to need you to help me with some 'creative accounting'.

The truth is that I've spent more time walking the river this year than fishing it. Even when I have caught a fish I've reeled in, sat back, and spent the remainder of the day drinking tea and watching the clouds float by. (Mrs H says I'm the most easily distracted person she knows. It's not entirely true. I like to observe the details around me, enjoying the nuances of the seasons and the changing light through the day, the 'always dramatic' British weather, and the opportunities to write about or draw things as they happen. So no. I rarely get properly distracted, just 'engrossed'. Ooh. Hang on. What were we talking about? Oh yes. I need your help to complete my catch return.)

We could go straight to the creative part, perhaps recording the fish in ounces or by the number of spots on their flanks? But I don't think that will be enough. I need your help, but only if you can pass a test. If you want the job, you'll need to read the following case study, answer four questions and then present the correct overall answer. You can use an abacus, calculator, fingers or toes, but be sure to record your answer, which you'll need to present as a one-hour slide presentation later on. Ready? Okay. Here goes:

The Case Study: Archie Wangumout goes fishing to catch fish. He fishes four times per week from April to October inclusive and two times per week for the remainder of the year. He catches, on average, five fish per visit and doesn't believe in catch and release. In the second week of May he fished a dry fly and caught twenty-nine trout. His January fish weighed 2lbs each but he noticed that the average weight of fish declined by two per cent per month, although his largest trout was an 8lb 4oz specimen caught in September. He fished a 9ft 6in rod with a 7-weight line for thirty-three per cent of the time, an 8ft rod with a 4-weight line for twenty per cent of the time and a 10ft lightning-hurler for the remainder. Archie is 47 years old, single, and his goal in life is to increase his catch rate by eighteen per cent.

Question One: Is Archie a total loser, or what?

Question Two: Will you give him a kicking, or should I?

Question Three: How long will it be before he gets sponsored for his efforts?

Question Four (the one requested by the Fishery Officer): What is the square root of the number of fish caught; added to the median weight of individual fish taken between May and September inclusive; divided by the average length of rod based upon time used; then multiplied by pie?

How did you get on? Did the case study make you sweat, curse and rub your brow, even before you'd seen the questions? Well done for persevering. I'll give you the answer later. For now, let's get back to the task at hand: debating why numbers are not important in fishing.

Recording numerical information is all very well and good for someone who counts beans, but not for an angler. Doing so only feeds the urge to tally away one's life, like a prisoner marking a cell wall. "He's a great angler," they will say, "He caught his thousandth salmon this week." Hmm. It depends what you mean by great. Or accomplished. Or contented. The fish, sadly, is just a number, added to a list of other numbers. We don't know what it looked like, how it fought, whether it was returned, or if the angler even smiled when he caught it. It is just number 1000, destined to be replaced by number 1001 and a whole load more until the angler reaches number 1037 and realises he's in serious need of meaningful entertainment. He'll then take up golf, or darts, or snooker, or any other sport where a number determines his enjoyment. The end result will still be the same: one-nil, where 'nil' stands for 'Never Investigated Life'.

It's the same with those who weigh or measure their catch. Again, the fish becomes a number:

"I caught six doubles this season."

"Interesting, I didn't know you watched tennis?"

"My average was eleven and a half."

"You mean eleven point five."

"Yes."

"I thought you had."

"Had what?"

"Lost the point."

As for the other so-called 'recordable details', like fly choice and weather, they at least document something that paints an image in one's memory.

It's a bit like knowing that a toad explodes at exactly 14 psi when you attach it to an air pump. But there are better ways of describing the seasons and documenting one's memories. I'll give you an example: think of your favourite holiday…

What came to mind? I'll wager that it wasn't 'atmospheric pressure 1013 millibars'? Most likely it was a sensation where you were happy, optimistic, excited and contented. All are feelings rather than, as asked, thinking; they are emotional states which made the holiday feel special and which, in turn, made it memorable. If angling is also special, then why think of it or record it in any other way than how it made us feel? For example, you've just been fishing and had the most wonderful day. But you caught nothing. Do you enter the word 'blank' in your diary? Or do you write something like, "Fished the Windrush – mayflies dancing in the evening sun – fledgling marsh tits twittering in the willows – primroses in flower – trout rising lazily – two token casts – happy just to sit and watch the sunset." As my friend Phin' says, "It's the anecdotes and incidents that make up a day. When we remember them, we're there, anecdotally and incidentally."

Here's the point that I most want to make: we are human beings who each experience our own journey through life, guided by our unique view of the world. Only we know our reality. Our perceptions, interpretations, feelings and thoughts are our own.

It's these feelings that get stored most vividly in our memory banks. One usually remembers how something made us feel, more than the details behind why we felt it.

So how do you feel about completing that catch return? The fishery manager will want us to report the facts – that Fennel caught three brown trout, each around the two-pounds mark. One on a Mayfly, one on a Goldhead Damsel and one on a Parachute Adams; two from the Windrush and one from the club lake. "But what was your catch ratio?" he will ask. "Three fish from how many trips?" The fact of the matter is I can't remember. It might be ten, or twenty, or thirty trips. Or it might be none. I might have made it up. I think I'll put "0.5" in the return and let him do the maths.

Postscript:
My apologies, I forgot to give you the answer to the fourth interview question. If you got "10.29" or indeed anything involving numbers, then I'm sorry, you haven't got the job. If, however, you made your calculations based upon Fennel's Principles of Numeric Reasoning then you'll have realised that there's a range of answers: chicken and mushroom, mince beef, steak and kidney, to name but a few. I'll go with steak and kidney, perhaps with some new potatoes and runner beans. That's my sort of pie, and my sort of memory.

Stop – Unplug – Escape – Enjoy

How do you measure the success
of your fishing seasons?

November

XX

LOOKING BACKWARDS, GOING FORWARD

Arnold the postman (who is far less glamorous than Beatrix von Baum, postwoman extraordinaire and star of my *Writer's Year* book) has just delivered the mail. An enticing package thumped onto the floor after it fell from the letterbox. I picked it up and found that it contained some early Christmas cards and several sales brochures. The latter were, it seemed, trying to convince me to buy everything from exotic holidays ("If I'd wanted to spend a week in *Tropical Itchybotty* then I'd have bought the cream in advance!") through to end-of-season 'never to be repeated' offers on sofas and beds. In amongst all this chaff was a catalogue from a mail-order fishing tackle company that I shall read while sipping a cup of tea. All plans are on hold while I put the kettle on and dream of new bits of tackle to add to my ever-growing (and mostly superfluous) collection. It's a programmed response, typical of most anglers. But I should know better. I used to work in the angling trade and know how the game's played. And herein is my confession: I once provided marketing and writing services to fisheries, bait manufacturers and tackle companies. Doing so taught me that angling is not the cottage industry it would lead us to believe. It's a *billion-pound* global industry. It wants us to spend money. Lots of it. It knows we love our hobby, that we yearn to shower our affections on fish by buying expensive tackle and – let's be honest – by looking the part. Wouldn't it be nice if we bought a new rod and reel every year, threw away our fly after catching a fish, bought a new coat every time it rained and owned gadgets designed for every angling situation? After all, it's not possible to catch a fish on a rod that's six inches longer than it should be, or on a hook that's already been blunted by a fish, or in a net that's designed for a different species, now is it? And what if you go fishing and the trout won't take any of the flies crammed into the twenty-six fly boxes concealed within your designer waistcoat? That just wouldn't do. You'll have to buy some more.

"Catching fish," the angling trade would have us believe, "has nothing to do with your angling ability or the appetite of the fish." If you find that you are forced to fish for up to twenty-two-and-a-half minutes without catching, then all you need to do is pop back to the tackle shop and buy some more gear. You'll then return to the water and catch a creel-full of fish within the hour. People will observe you in full flight, fully equipped to catch all and the largest fish in the water. They will yearn to be as good as you. "Hey Mister," they will say, "what rod are you using? I've never seen anyone cast like that. You're awesome!" With a confident smile and raised eyebrow, you'll look up from the water, dodge the fish that are throwing themselves into your landing net, and remark that it's the latest Tossworth 500 Uber-Action Airbender. "Unlock your potential," said the advert, "see and beat the horizon, go beyond the realms of the possible." This, of course, is bunkum. I should know. I was writing the marketing copy. But I'll be the first to confess to having tendencies towards 'angling image', needing to be appropriately dressed in tweeds and using vintage-styled tackle when fishing. I also believe that fishing, being my main hobby, is my opportunity to indulge and pamper myself as reward for all the effort and sacrifice I make elsewhere in my life.

Treating ourselves to the nicest wotnots and being able to display our angling preferences makes us feel good about our hobby and our self. But we have to be mindful of falling into the 'all the gear and no idea' trap. We must first work our angling apprenticeship before we can wholewalletedly fall into the marketer's trap.

Let's say you've made the decision to *really* get into fly-fishing. You've decided to invest in your hobby and purchase a quality fly rod. What would be a reasonable amount to spend? Six hundred pounds? That's two weeks wages for the average worker. But it's the current market price for a top-of-the-range trout fly rod. Let's say you save up, buy the rod and when you get it home you notice that there's a little certificate attached to it saying 'Unconditional Lifetime Guarantee'. Wow. That's impressive. You can stamp on the rod, crush it in a car door, or stuff it in a food mixer. If you break it, all you have to do is return the broken parts to the manufacturer along with a cheque for thirty pounds to cover 'administration fees'. They will send you a new rod, no questions asked. But wait a minute? What does this tell you? That the manufacturer can supply you a new rod, cover their warehouse costs and pay the courier – all for thirty pounds? But I thought the rod was worth six hundred pounds? Welcome to the world of mass-produced, branded fishing tackle.

Walk into a tackle shop today and it's likely that you'll only be presented with carbon rods. The angler rarely has a choice of rod material anymore. You might get lucky and find a fibreglass rod, but bamboo rods (wands of the gods) are sold by specialist craftsmen or in second-hand shops. The carbon cloud has overshadowed everything. Being a bamboo-boned traditionalist, I'm saddened by this, as I have always appreciated handmade things over factory-cloned items. I mostly use vintage bamboo rods. They help me to connect with a past era and a 'quieter' and slower way of fishing. But I am in the minority.

Nowadays most anglers use rods made from carbon fibre. Using one is like waggling a state-of-the-art tank aerial. They do connect, to lightning bolts and messages saying "Attack! Attack! Attack!" No wonder they make us want to cast faster and further into the distance? This high energy, high performance, slightly aggressive and domineering way of fishing – encouraged by the marketing man and the promises of 'unsurpassed performance' – is too macho for me. Fishing's supposed to be romantic, relaxing and leisurely, so why such eagerness? Young men love quickly; older men take their time. That's the truth about performance. And I've been fishing for a very long time. But 'performance' is not the only thing shaping rod technology and availability. It is also driven by cost.

We've already identified the sub-thirty-pounds manufacturing cost of a top-end carbon rod. The cheap production costs of carbon rods enable them to be retailed at prices to suit the whole market. This is good, as reducing the cost of entering the sport can only assist recruitment of new fishers to angling. A new split cane bamboo rod, on the other hand, can take up to eighty hours to build. Assuming a craftsman's wage of twenty pounds per hour and a retailer's mark-up of 100%, you can see why a modern bamboo fly rod can retail for more than £3,000. Aspirational ownership, yes (and one hell of a treat to yourself) but not for everyone's pocket. So you can see why carbon rods fill the racks of most fishing tackle shops. They are lighter, cheaper, perform better (in lengths over 8ft), require virtually no maintenance, and are entirely the right *rational* choice of rod to use. Their bamboo equivalents are heavy, expensive, slow-actioned, prone to damp and taking a 'set', but they're artisan-made, beautiful, lovely, gorgeous and entirely right for an emotional purchase. And because I love fishing, I will only allow myself to make *emotional* purchases.

Why am I harping on about carbon and cane rods? I'm not trying to sell you either. (Although, if you're interested, I have a dogleg, woodworm-riddled bamboo carpet beater in the shed that needs a good home. Yours for five hundred quid.) I'm trying to articulate that just because something's expensive doesn't automatically make it better. The latest technology isn't necessarily the best either. You can buy a cheap second-hand rod that will mean more to you than the glossiest 'new release' in a tackle shop window. With that in mind, where do these thoughts leave us at Christmastime when sales catalogues fall through the letterbox and we're thinking about next year's fishing? For me, it presents us with choice. If we want to spend a lot of money, or get the latest 'must-have' items of tackle, then so be it. It's our choice. If we want to keep using our old-but-loved tackle then that's fine too. Either way, it's our choice and not somebody else's. Which is just how it should be. Fishing is something we do in our free time, it's defined by freedom, so freedom of thought, opinion and choice is central to our sport. There is no right or wrong, just personal choice. Though, of course, it's all too easy to be brainwashed into thinking we need something, when really we don't. Be clear why you *want* something; never let 'needs' get in the way.

Postscript
It is said that the person we are today is the result of all the decisions we've made in our life. Even if someone has made a decision that influences us, it's our decision whether we accept how it impacts us or, more importantly, how it makes us feel. We have chosen to become the person we are today. Our personality and values are built upon our cumulative experiences and outlook on life, which is expressed in the choices we make. Knowing this makes us look backwards to understand the source of our identity. Then, when we're sure of the origins and meaning of who we are, we can confidently go forwards – with and towards things that have real value. This 'looking backwards, going forward' is the best way to avoid the marketing man who would otherwise tempt us with the latest and greatest, supported by an unlimited lifetime guarantee. We don't need the latest and greatest, unless we want it. First and foremost, we have our own limited lifetime to guarantee. How we spend it is our choice.

Stop – Unplug – Escape – Enjoy

What do you most want
from your fishing?

December

XXI

GRAYLING TO REMEMBER

Today is the 21st of December, the shortest day, the winter solstice, the runt of the year. A day so lazy it barely bothers to get out of bed. No sooner has the sun rubbed its eyes, stretched its rays and risen for its late morning brunch, then it's time for cocoa and an early night's sleep. It is a dull, dreary day that barely breaks free from the half-light. It buries its head in its hands at the prospect of winter, diving under a darkened duvet until spring.

At face value, today doesn't bode much promise for the angler. But that dusky light outside has got my angler's senses tingling. The air isn't especially warm outdoors, but I have a feeling that the fish could be feeding. The day may be only eight hours long, but it is an eight-hour 'witching hour' when my hunter's instinct is alert and my thoughts focus on the quest for fish.

The most appropriate fish to angle for at this time of year is the grayling, which should be in prime condition and hungry for a fly drifted past its nose. Something like a Red Tag or Grayling Witch or, if the river is high or coloured, a Czech Nymph bumped along the riverbed.

Maybe even a Killer Bug fished 'down and across'? Yes, a grayling it shall be. And I know just the venue.

The Gloucestershire Coln is the most famous of the Cotswold limestone streams. It rises in the hills east of Cheltenham, winds its way through some of the most picturesque Cotswold villages, such as Bibury and Coln St. Aldwyn, and then joins the River Thames at Lechlade. Famous angling writer Bernard Venables was the first to draw my attention to the river. His 1949 book *Fisherman's Testament* describes the Coln and the fishing it offers, concluding with the words: "The success of such days is not to be measured in catch alone. I had had a vastly happy day on a most beautiful water, where merely to be should be pleasure enough." I shall go fishing in an attempt to 'be' beside the river. And when I next write, it will be to update you about the adventure.

The Essence of Grayling

The weather was dull and cold when I arrived at the Coln. Drizzle was falling from the sky and the clouds were full of that amber tinge which made me think that snow could fall. I tackled up quickly and headed to my favourite spot on the river – a bend flanked with bulrush and overhung by an alder. The river here is narrow, just twenty-feet wide, so it provided intimate fishing where fish could be seen and stalked.

Just the sort of 'creepy-crawly' fishing I like. The beat of river is on a private estate, so it's neither trampled or overly-manicured. But it's by no means wild fishing, at least for the trout that are stocked each spring. It's a limestone stream, but not like the Hampshire chalkstreams; and it doesn't compare to the ruggedness and energy of upland rivers which usually require more than a thirty-yard walk from the car to reach the best spots. But it's local fishing, which has a charm all of its own. On a day like today, when the fishing window might only be a few hours, this river suits me fine.

Several grayling were rising just below the bend in the river when I arrived, but I couldn't detect what they were feeding on. (I could tell you that the flies were *Ecdyonurus dispar*, or *Soumatti meddup*, but I'm hopeless at entomology. Conclusion? 'Little brown things fluttering above the water'.) So I didn't worry too much about fly imitation and instead focused on getting my fly near to the rises. As Reg Righyni would have said, "I'd found the fish in feeding humour," so I'd have plenty of chances to change my fly if the first few drifts proved unsuccessful.

Casting a fly again (it is ten weeks since the trout season ended) felt pleasing but slightly awkward, like picking up a favourite fountain pen only to find the ink clogged in the nib. My first casts were sloppy and lacked conviction, as if my winter laziness had nudged my hunter's instincts aside in readiness for hibernation.

However, the life in the cane rod began to wave its magic over me. Soon I was fishing with fluid casts and a precision once described by a casting instructor as "Not bad, for an amateur." It was good enough for me. I cast just above the fish and allowed the fly to drift back towards them. The fly held in the surface film for a moment and then disappeared as the current pulled it under. It drifted about six feet before I noticed a slight bulge in the water and then the 'electric' tugging of a fish on the line. I lifted the rod and saw the line tighten, and then felt the unmistakable jagging, writhing, twisting fight of a grayling.

The battle was not as prolonged as I might have expected. The grayling, being hooked upstream, soon fell back in the current and was quickly guided into my net. I reached down and parted the mesh. A grayling. Oh how beautiful. Those pewter flanks and purplish-mauve fins. The large sail-like dorsal that has the iridescence of a peacock's feather, with greens and blues tipped with burnt crimson. And that unique eye, its pear-shaped pupil that reminds a traditional angler like me of Mr Crabtree's net.

Catching the fish reminded me of my favourite episode of *Angler's Corner* (a television series in the sixties) where Reg Righyni fished a northern river for grayling. Bernard Venables' softly spoken narration filled my ears and I could hear him saying, "The essence of grayling fishing" and "an essence of river mint and wild thyme." 'Essential' descriptions, but the water was too cold for my fish to have any noticeable scent. Bernard's commentary continued, "Grayling are a shoal fish; catch one then others will follow." Time, then, to return the fish and cast again.

I cast to the same spot as before and eventually caught another two grayling. "A brace and a half" – plenty enough for this angler, although I had no intention of taking the fish for the table. The river had been kind to me. I reeled in and went for a walk alongside the river, keen to warm my limbs and listen to the water. The sound of the river contrasted beautifully with the silence of the drizzle falling around me. It was like listening to water being poured into a church font at midnight.

The Perfect End

The time is now 4pm. The drizzle has ceased and the clouds are parting. I am sitting on an angler's bench, watching the sun set in the distance while I reflect upon an enjoyable day's fishing. I'm numb with cold, but I don't care. I'm sipping brandy from a hip flask and have charcoal-fuelled hand warmers in my pockets. I will stay here until dark, just long enough to demonstrate that my sitting here is as much a part of the day's activities as catching the fish. It is a token of respect for Bernard Venables and all the anglers who know that fishing is really about 'being' by water.

The grayling I caught earlier were merely my reason for coming here, not being here. My walk this afternoon was equally pleasurable (I had a robin land on my knee while I was sitting down and eating my sandwiches). Other anglers might think that I stopped fishing too soon, that my jottings were too brief and my walk was merely a way of staying warm on an otherwise soggy and freezing day. But it was real. Today is far from being a soggy cloth that needs wringing from my angling diary. It has been the perfect end to an angling year.

As I sit here and watch the light fade from the day, I see new memories form in my mind. I raise my hipflask and make a toast to *"Thymallus thymallus ut memor"*: 'Grayling, to remember.'

December

XXII

FINGERS AND FLUFF

I have in front of me a technical 'how to do it' guide to fly-tying. It's one of those spiral-bound books that flop open at just the right page for an expert and just the wrong page for a numpty like me. Whereas normal books encourage the reader to start at the beginning and read the chapter on 'How to Fix the Hook to your Fly-tying Vice', this book wants me to go straight to page ninety-six and get cracking on a fly that looks like it's crawled from the river and gone to sleep on the hook. The fly has got eyes, legs, a shell and gills. Crikey. The detail. I thought fishing flies were supposed to be made from feathers and fur and those silken tasselly things that you snip from the bottom of the sofa when your other half isn't looking? Oh, I understand now. This is a replica of *Dinocras cephalotes* and not *Perla bipunctata*. Of course, that explains everything.

I am officially out of my depth. I should have bought a book called *A Complete Beginner's Guide to Teaching the Hopeless Amateur how to Think About Watching Someone Tie a Fly, and if They Still Don't Get It then How to Take Them to A Shop and Buy One*. But given that this book would be incomplete without an article on fly tying, and given that I am a rank amateur at such things, I opted for a crash course in how to make a complete idiot of myself. Which is why I bought a book for the experts and why, after three hours of snapped silks, strained eyes and 'twisted tinsels', I am sitting here with a lap full of fluff and a headache that would put the consequences of a swiftly drunk bottle of port to shame. It's time, if I can muster the mental energy, for a rethink.

I have a dream that one day I'll be able to tie perfect examples of such evocative flies as the Wickham's Fancy, Adam's Irresistible and Tup's Indispensible; and that I might eventually get radical and attempt the intriguingly-named 'Thunder Chicken' that I read about on an American fly-tying website. At some point in the future I'd like to head off to Cumbria or the Scottish Highlands with nothing but a rucksack, my fishing tackle and a fly-tying box to indulge the holidaying whims of the romantic fly fisher. Alas, such ambitions are beyond my current reach. Whilst I might be able to tie crude flies from the contents of a dustbin, anything requiring a hackle, legs or wings ends up looking like Worzel Gummage's toothbrush or, worse, something you might find in his belly button. I therefore confess to lacking the necessary expertise to talk to you with any authority about tying flies. But I know a man who can.

My father is the person who introduced me to angling, just like his father introduced him.

FINGERS AND FLUFF

That's how it goes in a fishing family. My early exposures, by today's standards, were a little extreme. As a toddler I'd accompany Dad fishing and, for safety's sake, he'd wrap a rope around me and then tie me to a stake in the ground while he waded in the river. As I grew older, and learned how to wriggle free from a Bowline knot, I was allowed to carry Dad's creel and help him choose flies. Then, at the age of five, I was allowed to become an angler in my own right and pose for the picture that inspired this book.

Dad is a fly fisherman. He began tying his own flies in the mid-1970s and has amassed more knowledge and experience on the subject than I could ever hope to achieve. I'm going to ask him to describe a very special fly called the Hudson Sedgetastic, a deer hair fly that has caught more trout than any other for my family. You've probably heard of the angling gene? Well in our family we have an angling legacy: a fly that can be relied upon more than any other, a pattern that is handed down from generation to generation and with it all the achievements of the past and the dreams of the future.

Dad, over to you:

Thanks, Fennel. The Sedgetastic is a variant of the 'G&H Deer Hair Sedge' that I first tied in 1977 after reading 'The Super Flies of Still Water' by John Goddard. It differs from the original by having a hackle trimmed underneath rather than above, not having a seal's fur underbody and omits the two 'antennae' near the eye of the hook *(which frequently caused tangles when casting)*.

The fly and its colour variations have contributed to at least eighty per cent of my catches over the past thirty-two years. Relatives and friends report similar success. Consequently, I have greatest confidence in this fly, I fish it more often than any other and rely upon it in all but the most severe of conditions. I have caught throughout the year on the Sedgetastic, on all waters – be they stocked or wild fisheries. Some anglers claim that the fly represents nothing but a trout pellet. This is clearly not the case as I have caught many wild trout and coarse fish (a carp on one occasion) that have never seen anything other than natural food.

Many times I have been asked by my neighbouring angler what fly I am using when I am catching and he is not. On producing the Sedgetastic I have often been met with derision and laughter. "You're having me on!" is a frequent response. I always give the doubter one of my spares and in more cases than I can remember the angler has had an almost instant take from a fish. They then say, "But I can't tie a fly; where can I buy it?" or "What's the dressing; how do I tie it?" (And sometimes the angler will say, "Thank you very much!")

Like many dry flies, the Sedgetastic will attract trout when it lies motionless on the surface or drifting in the wind; or it can be gently tweaked, pulled slowly or stripped back. It is a great bob fly on a cast (though I rarely use more than one fly as I get in such tangles) and it can be used as a general pattern *'upstream dry' when fishing rivers. It often provokes a rise on seemingly 'dead' water when no other rises are visible. During dusk and into dark it can produce savage takes when stripped back.*

Due to its deer hair body, the Sedgetastic is very buoyant. It can be fished high in the water when dry, or in or just below the surface when waterlogged. I've even fished it 'booby-style' on a sunken line. However, I normally fish a floating line with a subsurface cast. Do I use floatant on the fly? I do not. Frequent false casting is enough to dry the fly and keep it floating.

My advice is not to look at the artificial as you see it but to imagine how a trout sees it from underneath. What does it look like on the surface with the sky as the background? It's not a lure: it has the shape and character of a food source that looks interesting and worth tasting. It's a scruffy fly that keeps on catching, even after it's been chomped by a fish.

Materials for the Sedgetastic

Hook: 10 or 12 long shank, light wire.
Thread: Black.
Body: Deer hair, spun and trimmed to shape. Natural deer hair is standard, but other colours produce results: white deer hair is great at dusk as it helps make the fly more visible (and, I believe, makes it look like a moth); black deer hair (either shop bought or dyed with a permanent

marker pen) is good; you can even use fluorescent deer hair when you want a rise from your fellow anglers as well as the trout. (Just colour it black underneath with a marker pen so the fish don't know any difference.)

Hackle: Two red cock or two badger cock.

How to tie the Sedgetastic

1. Wind thread down the shank of the hook.

2. Tie-in a pinch of deer hair with two turns of thread, then pull the thread to spin the hair. Add a turn of thread to the hook to secure.

3. Repeat Stage 2 up to four times, butting up each bunch of deer hair (not too tightly) until you've worked along the shank of the hook. (Remember to keep enough space near to the eye for the hackle.) The fly should now look like an unkempt hedgehog.

4. Tie off the thread with a whip finish.

5. Using scissors, trim the deer hair to a slim wedge shape. Cut the hair flat (square-edged) under the shank and quite close to the wire. This helps the fly to sit nicely on the surface and improves hooking.

6. Tie in two cock hackles and make three or four turns. Tie off with a whip finish.

7. Trim the bottom of the hackles to a flat profile. This helps the fly sit in the surface film without keeling over.

That's it. The Sedgetastic is a great fly. It's not the easiest or quickest to tie, but it's worth the effort.

FINGERS AND FLUFF

If I were to have only one fly in my box, then this would be it. It catches trout. Lots of them. The question is, are you willing to try it?

Thanks, Dad. I knew you'd be able to help. I'll conclude with the advice given to me by an elderly gentleman I met while out walking. He was sitting in a deck chair in his front garden, with a cigar in one hand and a cognac in the other. "Young man," he said, "I've got the decorator's in. If there's one thing I've learned in life it's this: if at first you don't succeed, get someone else to do it."

Which reminds me: Dad, any chance you could make me some more flies?

> *Stop – Unplug – Escape – Enjoy*
>
> What's the go-to fly in your box?

December

XXIII

THE FOUR RULES OF ANGLING

Thirty years have passed since I became an angler. From my first cast into a mountain reservoir in Wales, to the rivers of southern England and to my inevitable return to the Welsh mountains, angling has been my one reliable source of relaxation and the one thing that has remained true and honest. It's been the consistent thread that I can follow to find my way back to my safest place. Over these thirty years I've sought many things from angling. At first it was the excitement of catching a fish – any fish – then it was the joy of catching lots of fish, then the skill of knowing how to catch a chosen species and finally (or so I thought) the ability to catch large fish. I began by blindly casting a spinner into open water; then came the joy of rising a trout to a dry fly; then came too many tactics and techniques born of an obsession with fishing.

I regard the angling I did during childhood as the most special. I knew nothing of angling 'labels' or politics. There was no such thing as game or coarse fish in my vocabulary. A fish was a fish. I would fly-fish, leger a worm, spin or tickle my trout;

I then enjoyed catching perch, roach, dace, chub, carp, tench, bream, barbel and pike. I was entirely catholic in my tastes and not swayed by rules and etiquette. But then, as I got older and sought bigger fish and more exclusive fishing, things became stuffier and rigid. I liked this at first, but eventually felt the need to 'fly' free from these waters to rediscover the simple pleasures of angling. I retraced my footsteps to find a junction that might lead me to a different and more fulfilling future. I fished progressively wilder waters for smaller and prettier fish; I carried less and less tackle and sought to catch fewer and fewer fish. And then, while lying upon a riverbank and looking up at the sky with a childhood friend, found the junction I needed when he said: "Y'know, Fennel, there are four rules of angling. First: all fish are equal, and all are special. Second: never cast before the time is right (and only you know when this time shall be). Third: be a gentleman angler; be honest, considerate and kind; riches come not from receiving but from giving. Fourth: always follow the smoke from your pipe, because you never know where it might lead."

I reflected upon my friend's words. After much soul searching, I concluded that what I seek from fishing is not a fish, or merely the chance to escape, rather it is the joy I felt at the age of five when I held a beautiful little trout in my hands and wept because the world, as I saw it, was so perfect.

# ABOUT THE AUTHOR

FENNEL HUDSON

"Author, artist, naturalist and countryman. His is a lifestyle to inspire the most bricked-up townie."

Fennel Hudson is a lifestyle and countryside author known for his *Fennel's Journal* books, *Contented Countryman* podcasts and *Flyfishers' Journal* articles. A champion of fishing for wild fish in remote and tranquil locations, he's an angler who's more likely to be seen wearing hiking boots than waders. He travels light and fishes simply, always seeking an adventure into the natural world: the very thing that makes fly-fishing so appealing. His quest is as much for freedom as it is for fish. When alone upon a mountain, or beside a stream, he cannot help but 'Stop – Unplug – Escape – Enjoy'.

For more information please visit:
www.fennelspriory.com

THE FENNEL'S JOURNAL SERIES

THE FIRST-EVER REVIEWS OF FENNEL'S JOURNAL:

"Fennel's Journal began as a series of illustrated letters to friends. As these evolved they became less a diary, more a manifesto, and the Journal is now exactly that – a way of living, rurally and simply: very real for all those who recognise the importance of tradition and joy."

Caught by the River

"I can see where it might lead. What he has would make amazing TV. It's the Good Life, but in a realistic way. It's Jack Hargreaves. It's Countryfile. It's quality Sunday newspaper stuff. It's 1948, all over again. In trying to escape the present he's inevitably created a brand. A potentially very powerful brand."

Bob Roberts Online

"Fennel's Journal is a masterpiece about rural living. It is a route-map to the life we all seek."

The Traditional Fisherman's Forum

From A Meaningful Life:

"Life is the most beautiful and rewarding gift. We just need to take time out to allow us to reflect, change perspective, and see things in their best light. Sometimes we just have to stop and feel the pulse of the Earth, the rhythm of the seasons and the internal voice that was once our childhood friend. As the natural world grows smaller, so too does its intensity and the size of the window through which it may be viewed."

NO.1

A MEANINGFUL LIFE

A Meaningful Life is the first and perhaps most important Journal. It documents the origins of Fennel's Priory and why Fennel decided to live by a new set of ideals. With themes ranging from escapism, adventure, work-life balance, identity and purpose, through to traditionalism and country living, it sets the scene for future editions – building messages that are central to Fennel's Priory. Ultimately it conveys the importance of a relaxed, balanced, and meaningful life.

READER TESTIMONIALS

"I loved reading this Journal. It's inspiring and has the beginnings of something very special."

"Fennel's chosen trajectory is firmly in the slow lane. He's a countryman, with courage to stand behind his traditional values."

"Witty and emotive, Fennel's writing conveys passion for a slower-paced and quieter life."

From A Waterside Year:

"Water is intrinsically linked to the mystery and excitement of discovering new worlds. Of dreams. And hopes. And thoughts of what 'could be'. Dreams free us from normality. ...As the daydreams grew longer, the distinction between what was real and what was imaginary grew less. Soon I existed in a blissful world of my own creation. Reality, as I learned, is only a matter of perception...A life that is real to one is surreal to another."

NO. 2

A WATERSIDE YEAR

In *A Waterside Year*, Fennel takes time out to live beside a lake in rural England. Here he appreciates the healing qualities of water, studies the wildlife around him, lives at the pace of someone outside of normal daily life, and discovers the freedom that's found in isolation. Getting so close to Nature, and spending time in idle fashion, enables him to discover a stronger sense of self. Ultimately he learns that freedom is not a place, but something that exists within us.

READER TESTIMONIALS

"A year in the wild. How we would all love to follow in Fennel's stead and indulge our dreams, to come out the other side a stronger and wiser person."

"A Journal with a message – that we should take time out to think about what's important, and see the beauty of the world."

"A truly blissful read full of inspiration and humour. The story of Fennel sitting in his tent, with the noises outside, had me laughing out loud!"

From A Writer's Year:

"Writing, with a fountain pen and ink from a bottle, is the simplest of things. Yet it can transport us to a different place entirely. Imagination is the real magic that exists in this world. Look inwards, to see outwards. And capture it in writing."

NO. 3

A WRITER'S YEAR

A Writer's Year celebrates the writer's craft. It champions the handwritten letter, discusses vintage pens and writing ink, and celebrates things such as antique typewriters and the quirkiness of the creative mind. It's a blend of observations. It's funny. It's serious. It's real life. But most of all it is written to encourage aspiring authors to find their voice, to put pen to paper, and follow their dreams.

READER TESTIMONIALS

"Worth it for the first chapter alone. It cannot fail to motivate and inspire the would-be author."

"What Fennel has written is not so much a eulogy for the handwritten letter as a call-to-arms for everyone to follow their dreams and make the most of their God-given talents. This is a genuinely inspiring read."

"I loved the part: 'If a pen can communicate our thoughts, dreams and emotions and be the voice of our soul, then ink is the medium that carries the message'. It shows how important and generous writing can be."

From Wild Carp:

"Some will say that searching for your dreams is like looking for unicorns in an emerald forest. They will say that following a golden thread will lead only to a king, dethroned and living in the gutter. This may be so.
But the king was made, not born.
The crown was never his to wear.
...If ever the adventure proves tiring, or you lose sight of your dream, look to the west at sunset. There, on days when the skies are clear, you might see upon the horizon a thin layer of golden mist. When it appears, you will know its purpose: it is the mist of believing."

NO. 4

WILD CARP

Angling for wild carp is about adventure, history, atmosphere and emotion. *Wild Carp* captures this aplenty, describing Fennel's 20-year quest to find a very special type of fish. But it's also about nature connection and a desire to uncover the seemingly impossible – a place where we can discover and live out our dreams, to completely indulge the mantra of 'Stop – Unplug – Escape – Enjoy'.

READER TESTIMONIALS

"When written well, traditional angling writing by the likes of BB, for example, is the type of literature that I can read again and again. Fennel's writing flows un-hurried without overly romanticising each point and the research is thorough; from the first sentence I was thinking, 'this lad can write!' It's informative and very refreshing."

"Such inspiring writing. His words 'Somewhere in the undergrowth of the impossible' had me staring out from the page in amazement. Fennel's writing is pure poetry."

From Fly Fishing:

"The deeper we travel into the natural world, and the greater the number of technological encumbrances we leave behind, the more likely we are to escape the fast-paced lifestyle and stresses of the 21st Century.
For some, angling enables a quest into the unknown, an adventure into the wild. For these fortunate folk, fly-fishing is escapism. Their hours by water serve as contemplation to enrich their souls, directing their quest inwards, towards their longed-for state of completeness."

NO. 5

FLY FISHING

Fly Fishing celebrates the most graceful and artful form of angling, explaining what it means to be an angler – in the spirit of Izaak Walton – and how fly fishers differ from bait fishers. The sporting and aesthetic beauty of fly-fishing is described in Fennel's usual witty and contemplative style. As he says, "Fly fishing is the ultimate form of angling; it gives us a reason to fish simply, travel lightly, and explore wild places that replenish our soul. With a fly rod, we're not casting to a fish; rather to a circle of dreams: ripples that spread into every aspect of our lives".

READER TESTIMONIALS

"Brilliant writing. Fennel made me laugh out loud in bed. My wife was asking questions!"

"A delightful, well-articulated, read. I strongly recommend it, especially to the contemplative, tradition-loving, bamboo fly rod devotees among us."

"A very inspiring and rewarding read. I will try to tie the Sedgetastic fly. It looks tasty!"

From Traditional Angling:

"Physics teaches us that for every action, there is an equal and opposite reaction: a natural balance of energy that sustains the equilibrium of life. In modern angling, these forces are skewed so far in favour of technology that the balance between science and art has been lost. But there is a movement, an undercurrent that defies the flow of progress. There are those who choose not to follow the crowd. They seek not to fish in a predictable, scientific manner. They yearn for the opposite, to buck the trend, *to be different*. They are the Traditional Anglers."

NO. 6

TRADITIONAL ANGLING

Traditional Angling celebrates the Waltonian values of angling: about fishing in a seasonal and uncompetitive way for the pure pleasure of being beside water. It wears its heart on its sleeve and a wildflower in its lapel. It's passionate, provocative and eccentric, written for those who appreciate the aesthetics of angling and uphold its sporting traditions. So, with great enthusiasm, raise your bamboo rod aloft for an adventure that proves there's more to fishing than catching fish.

READER TESTIMONIALS

"A beautifully written, very engaging and hugely enjoyable read. In fact, it's the best thing on fishing I've read in a long time."

"What a Journal! Fennel is clearly the spiritual successor to his mentor – the great Bernard Venables. There's so much wisdom and craftsmanship in his writing. Bernard clearly taught him very well."

From The Quiet Fields:

"The countryside, with its vast horizons, fresh air and ever-changing seasons is, by its very nature, more life-giving and adventurous than any amount of modern indoor living. It inspires a love of natural history – everything from the birds that sing in the trees to the quality and richness of the soil beneath our feet. Most of all, it creates the desire to exist more naturally. And in doing so, we appreciate the balance of life."

NO. 7

THE QUIET FIELDS

The Quiet Fields is rooted in the humus-rich soil of the countryside. It's about remote rural places where Nature exists undisturbed, where we may sit and ponder 'The Wonder of the World'. The Journal tips its hat to these places, and to the nature writing of BB, revealing the 'Lost England' that still exists if you know where and how to look. It is the most sentimental and astutely observed Journal to date, discussing the 'true beauty' of Nature. If you've ever yearned to hear birdsong during a busy day, then this is the book for you.

READER TESTIMONIALS

"Fennel's writing reminds me of the works of Roger Deakin. It inspires me with faith in the quiet life and that although I may be isolated, I am certainly not alone."

"Fennel has captured the essence of the countryside – that is, its almost human character. So brilliantly has he compared and contrasted it with the nature of we humans. It's not so much a 'balanced study', more a 'study of the balance' between Nature and Man."

From Fine Things:

"It seems that, depending upon which side of the thesaurus-writer's gaze we sit, one's uniqueness as a person can be deemed to be either eccentric or distinctive. Both, in my opinion, are good...As we get older, and experience more things, those of us with strength of character and a sense of purpose will grow stronger and fight harder; those who lack identity and direction might end up sitting in a corner somewhere, blindly taking all the knocks that life throws at them. What does this teach us? That character and purpose are directly linked to confidence and conviction. What links them? Courage – to be oneself, no matter what others might say."

NO. 8

FINE THINGS

Fine Things celebrates the special and sentimental items and activities that convey our personality. The writing is fast-paced, quirky and humorous, reflecting the author's enthusiasm and eccentric view of the world. But be warned: if you look inside Fennel's mind, you might see a hula-hooping hamster named Gerald, shaking his maracas, loudly banging a bongo, and getting him into all sorts of trouble. So strap yourself in. This book picks up pace and takes some unexpected turns. From the deeply personal to the outright eccentric, it's for those who seek to be different.

READER TESTIMONIALS

"A very fine thing, indeed. Fennel's best and funniest book to date. He is the only author who can make me laugh out loud and cry in the same sentence. I was constantly in tears, for all the right reasons."

"Deep in places, outright bonkers in others. A demonstration of the fine line between genius and madness."

From A Gardener's Year:

"Roll up your sleeves and imagine your vision of paradise. This, in whatever form it takes, is your garden. Keep hold of the image; know it's every detail and piece together the elements that need creating or nurturing, so that when you get the chance, you can prepare the ground, sow the seeds, and make it real. Ours is a gardener's life, whether we realise it or not."

NO. 9

A GARDENER'S YEAR

A Gardener's Year celebrates the joy of growing things and reflects upon a life working with plants. But it's not a record of horticultural activities through the seasons. It's a metaphor for having a dream and making it come true. For Fennel, who has spent half his life working in gardens, it's about cultivating a cottage garden where he can aspire to a self-sufficient lifestyle. The Journal sees him sow the seeds of this future reality.

READER TESTIMONIALS

"Fennel's writing is uniquely funny. I mean, who else can name a chapter 'Chicken Poo'? His sense of humour, balanced with some deep yet subtle messages, had me in tears. From his 'escape' to a public toilet, to what not to say to a celebrity, this is a Journal to entertain all readers."

"When I started reading this Journal I had a garden with a lawn and a patio. Now I have a vegetable patch, blisters, an aching back, and the biggest smile of my life. Thank you Fennel!"

From The Lighter Side:

"If self-actualisation is the pinnacle of one's development, then it can't be achieved if your mountain has two peaks...Being the 'best version' of yourself implies that you have other versions kept locked in a closet. Don't have any 'versions'. Just have one true, beautiful and pure form of you.
So climb your mountain, open your arms to the Creator who greets you there, and sing loudly to the world that stretches out beneath you. Write your name permanently on the landscape of your mind.
Remember: you are a child of Nature. And you are free."

NO. 10

THE LIGHTER SIDE

There's a delicate balance between something meaning a great deal and that same thing becoming so serious that it's ludicrous. (Ever got stressed about what clothes to wear for an interview?) That's why *The Lighter Side* provides the encouragement, humour, anecdotes, reflections and honesty that are essential to Fennel's message of 'Stop – Unplug – Escape – Enjoy'. After all, we can only 'Enjoy' if we know how to smile when we get there.

READER TESTIMONIALS

"The Lighter Side was more than I expected. The deeper meaning within it – and the devastating honesty it conveys – made me question exactly where I am in my own life and what I can do to improve it for my family and me in the time that remains. Thank you Fennel for opening my eyes and adjusting my course."

"The opening chapter is the most startling, erudite, compassionate and open piece of writing I have ever read…thank you Fennel for sharing so much. It did and does mean a great deal."

From Friendship:

"What I'm talking about is proper friendship. The sort that is authentic, genuine and real. Where we can look into the eyes of another person and know what they're thinking. ...Because, as friends, we remember 'why' as much as 'when' or 'what'. Through good times and bad, we were there. Together. That's the bond, the unquestionable obligation that's freely given. It's the tightest hug, the biggest kiss, the tearful hello and the widest smile. If that's what it means to be a friend, or an extrovert, or just someone who cares for others then that's me to the last beat of my heart."

NO. 11

FRIENDSHIP

Written by the Friends of the Priory, with bonus chapters from Fennel, *Friendship* provides insights into what it means to be friends, how shared interests and beliefs support collective purpose, and how, when we're together, we can achieve more, appreciate more, and have more fun. It's about the broader world of Fennel's Priory and how it exists in others. It's a book written 'for us by us', with friendship as the theme.

READER TESTIMONIALS

"Possibly the greatest gift that this Journal bestows is to let us know that we are not alone."

"Like friendship itself, this Journal brings together people and meaning. It reminds us that 'together we are strong'. Thank you Fennel for leading our charge."

"The message (and evolution) of Fennel's Journal is most evident in this Friendship edition. With such obvious themes as identity and legacy, it's clear that what Fennel has shared over the years is a route-map to freedom and a stronger sense of self."

From Nature Escape:

"I am once again seeking an escape, to where I hope to find freedom and connect with the young man who handed me his trust ten years ago. This will be a faithful interpretation of the Priory, a fitting way to mark ten years of writing. As I said at the end of last year's Journal, 'One's journey through life is not linear; it's circular.' So let's go back to the beginning, and rediscover the quiet world."

NO. 12

NATURE ESCAPE

Nature Escape provides the most detailed account of a day that follows the motto of 'Stop – Unplug – Escape – Enjoy'. In it Fennel returns to the woodland of his youth to study its wildlife and savour its peacefulness.

Written in real-time, with twenty-four chapters that each represent an hour, the Journal is an account of how time spent outdoors in wild places enables us to observe the nature that's around us *and* within us.

READER TESTIMONIALS

"Fennel's Journal has always provided us with an escape, but now we know where the escape can lead. As promised, it leads to enjoyment – and very enjoyable it is too!"

"24 hours alone in a wood, with only 'the wild' for company? With Fennel as our guide, there's no such thing as 'alone'; only the warmth of knowing that quiet times are the fine times."

"By studying the nature within us and around us, Fennel demonstrates how to be 'at one' with nature."

From Book of Secrets:

"There's a greater man than me who can sum up our journey, a mountaineer who in 1865 first climbed the Matterhorn. Edward Whymper, over to you: 'There have been joys too great to be described in words, and there have been griefs upon which I have not dared to dwell, and with these in mind I say, climb if you will, but remember that courage and strength are naught without prudence, and that a momentary negligence may destroy the happiness of a lifetime. Do nothing in haste, look well to each step, and from the beginning think what may be the end.'"

NO. 13

BOOK OF SECRETS

Book of Secrets links all editions of Fennel's Journal together. With 14 Journals in the series, and 14 core chapters in this book, it's the 'one book to bind them all' with each chapter providing the continuity story from one Journal to the next.

Containing Fennel's previously private writing, it provides deep insight into the Fennel's Journal story. If you've ever wondered why each Journal is themed the way it is, or tried to find the metaphor in each edition, then *Book of Secrets* is for you.

READER TESTIMONIALS

"What a privilege: being able to read the private writing of my favourite author. Book of Secrets is a treat."

"Such honesty and wit. Fennel puts into words what I have only ever thought, or dare not say."

"Fennel's Journal really is a series – it's meant to be read as a whole. And now we have the key to unlock it."

From The Pursuit of Life:

"We can hide, or we can strive – for a life of our making. With endless possibilities and opportunities to reach for our dreams, we owe it to ourselves to dream big and keep going, irrespective of what we might encounter. Sadly, the thing that most limits our success is not others, but ourselves. How strongly we believe, how confidently we act, how fiercely we react, how passionately we want, and how life-affirmingly compelled we are to grow and blossom; that's how we keep going, no matter what, to be the person we want to be, living the life we deserve, in dreams that are real."

NO. 14

THE PURSUIT OF LIFE

The Pursuit of Life concludes the Fennel's Journal story. It's a reflective tome that provides Fennel's commentary on the journey and a 'behind the scenes' view of the challenges and rewards of a life rebuilt on one's terms.

It's an account of how the series came to be and how it evolved, and includes much of Fennel's private writing, several of the original handwritten drafts, correspondence between The Friends, and encouragement for those on similar paths. Ultimately it shows how the Fennel's Journal series can be used as a route map to a more fulfilling life.

READER TESTIMONIALS

"A life retold, for our benefit. Fennel is to be congratulated for everything he's achieved – on paper and in life."

"It's his life in the books, but it could so very easily be ours. Fennel has a way of seeing truth in the severe and the sublime, and bringing it home."

"Can this really be the end? When dreams are real, we never wake from them. More books Fennel, please!"

www.ingramcontent.com/pod-product-compliance
Lightning Source LLC
Chambersburg PA
CBHW030331230426
43661CB00032B/1369/J